Better Posters

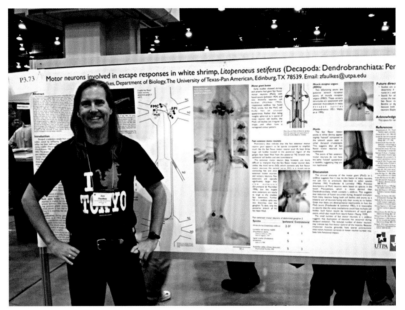

The author in his natural habitat: presenting a poster at a conference.

Better Posters

Plan, Design, and Present
a Better Academic Poster

ZEN FAULKES

PELAGIC PUBLISHING

Published by Pelagic Publishing
PO Box 874
Exeter
EX3 9BR
UK

www.pelagicpublishing.com

Better Posters: Plan, Design, and Present a Better Academic Poster

ISBN 978-1-78427-235-7 Paperback
ISBN 978-1-78427-236-4 ePub
ISBN 978-1-78427-237-1 PDF

A CIP record for this book is available from the British Library

This manuscript was written in Microsoft Word (www.office.com).
Figures were made using Origin (www.originlab.com), CorelDRAW,
and Corel Photo-Paint (corel.com). The "hand-lettered" typeface used
in many figures is Unmasked from Blambot Studios (blambot.com).

For my family: my wife Sakshi, my pack Max,
my father Kevin and my mother Karren

Contents

Part III For organizers

Part IV What next?

Foreword

If you're like most academics, you have not received formal training in verbal or visual communication skills. So, it's not all that surprising we frequently commiserate about jargon-filled wall-of-text conference posters and #DeathByPowerPoint presentations.

It seems that most (if not all) of us know there is a problem with how academia communicates their research and educational material. Luckily, there is a group of folks dedicated to fixing this problem and helping academics more effectively communicate their work. Zen Faulkes (or, as he is also known on Twitter, Doctor Zen) is one such person that has been helping academics create more engaging and effective conference posters for years. It's why I am so excited for Doctor Zen's book on poster design and was happy to write this foreword for this book.

I train educators (e.g., academics, scientists, researchers, evaluators) to create visually engaging and effective slide presentations. In other words, my focus area is on ending #DeathByPowerPoint. There is, however, a lot of overlap between visual presentations and conference posters, and almost all the people I work with create both posters and presentations. So, I keep an eye out for poster design resources that I can share. Doctor Zen's poster design website was one that I often shared with others.

I published my first blog post about posters because I kept getting requests for templates, and even saw conferences starting to recommend or require templates. As a presentation designer and trainer, the number one biggest struggle for me is convincing people that a slide template will not solve their problems. I've lost count of the number of times folks have asked me to design a template for them, in hopes that this template will make their presentations visually engaging from that point on. It's taken years of educational work to explain to folks that #DeathByPowerPoint is caused by something much deeper than the wrong template. The last thing I wanted was for the idea of a template to become expected (or even more expected) in the poster design field, too.

The blog post I wrote about why templates aren't the solution for posters (or presentations) caught Doctor Zen's attention – that's how we connected, and ultimately why I'm writing this foreword. In my opinion, this book fills a huge gap in the training literature on how to create an effective conference poster. Many of the folks I work with have asked for help with applying the presentation design strategies I teach to their conference posters. I have a long to-do list, and one of the items (at the bottom) was to create lessons about poster design for people. Now that this book is available, I can cross that item off my list and recommend this book instead. It's that comprehensive, and it aligns with the design principles and critical thinking skills I teach in my presentation design training.

What I appreciate most about Doctor Zen's poster design book is that he effectively explains why templates are not the solution for poster design (which can be directly applied to presentation design!). He argues that academics and scientists should learn design thinking and graphic design skills as a foundation for knowing how to design a poster. To my delight, he goes beyond that and also talks about ways to emotionally resonate with people instead of just throwing a giant wall of facts and data at them. These are the principles that will help you design conference posters that make an impact and will be more likely to catch people's attention, help them understand your content, remember your content, and (ideally) use it later on. That's the power of effective communication, and as you will soon see, why we need to go beyond superficial template designs. You will find an excellent book to help you learn practical design thinking and graphic design skills. The examples and visuals provided throughout are helpful, with just enough humor to make this a fun read as well.

A pleasant surprise was finding out that this book goes beyond poster design tips. Doctor Zen also explains the context of a poster session and conference more broadly. If you are a first-generation college student like I am (or if you've never attended a poster session) then you will find this additional information to be a valuable resource. This is the book I wish I had had before going to my first poster session or creating my first poster as an undergrad student. Back then I had no idea what to do, what the audience was expecting, or what poster presenters were expecting of me. I was overwhelmed, confused, intimidated, and ended up glancing at a few things and leaving because I didn't know the social code of conduct for poster sessions. Doctor Zen shares information about that, as well as important conference basics such as the difference between posters and other common conference formats. Again, as a first-generation student, I didn't understand what most of these terms meant when I applied to my first few conferences.

Overall, Doctor Zen has provided a valuable contribution to the academic field with this book. Thank you for picking up this book, and thank you for working to make your poster more effective. If we see more people following the advice in this book, and fewer people searching for a template to solve their problems, then I see an end to jargon-filled, visually starved, wall-of-text conference posters. Instead, I see a future of poster sessions filled with excellent design, creativity and – most importantly – with folks who are able to communicate their work effectively in ways that make a lasting impact.

Echo Rivera

Dr. Echo Rivera is the owner and founder of Creative Research Communications, LLC, a Denver-based company that specializes in graphic design consultancy for academics. Her website is www.echorivera.com and she is @echoechoR on Twitter and Instagram.

Preface

I created the Better Posters blog (betterposters.blogspot.com) in March 2009 out of self-preservation. I'd seen so many bad conference posters, and made more than a few myself, that I hoped that if I blogged about them there might be a little less ugliness in the world. Plus, I had noticed that there was an unfilled niche for discussion about poster design for academics. There were static websites with good advice, but nothing that was continuously updated. It felt like people thought conference posters were a solved problem with nothing left to say about it, but I thought there was so much more to discuss. I wrote the blog for fun and for free because I've always believed academics were public figures who should try to make their ideas free.

But something surprising happened: people started reading the blog. (This is never a given for anything online.) Even more surprising was that people started sending me their posters, let me show them on the blog and criticize them. And as the blog grew in popularity, people started to tell me that they were recommending the blog to people, particularly to students who were about to make their first poster. I appreciated the recommendations, but increasingly I realized it was unfair to expect anyone making a poster for the first time to trawl through years of weekly blog posts that were posted with no plan beyond "I think I'll write about *this* today." (This is the downside to having a long-running and ongoing project.)

So I wrote this book because there was a need for something different than the blog. People needed a something that was more like a start-to-finish poster-making manual that was coherently organized. The ideas are still free and will continue to be explored on the blog, but this book can provide a key, a guided entry ramp, into the process of making posters in a way the blog does not. I hope it helps.

Acknowledgments

Thanks to Gavin Abercrombie, Emily Austen, Spenser Babb-Biernacki, Craig Bennett, Mary Bratsch-Hines, Veronika Cheplygina, Neil Cohn (www.visuallanguagelab.com), Giovanni Dall'Olio, Jacquelyn Gill, Kayla Hall, Kristina Killgrove, Milan Klöwer, Rajika (Reggie) Kuruwita, Cheryl Lantz, Mike Morrison, James O'Hanlon, Ana Maria Porras, Martin Rolfs, Kelsi Rutledge, Jessica Schubert, Jessica Stanton, Nicholas Wu, and Charles Cong Xu for giving me permission to use their material in this book. (This is why I love academics: they will do a lot of things if you ask just because they want to see a little less ignorance in the world.)

Thanks to the many contributors to the Better Posters blog for always giving me something to write about. The blog wouldn't be what it is without them, and this book probably would not exist without them.

Liz Neeley was one of the first people to recommend the poster blog. Her plug made me think I was on a good track and I am forever grateful.

I thank my department chairs Frederic Zaidan III and Kristine Lowe for giving me teaching schedules that allowed me to pursue writing this book.

I thank the town of Saint Johnsbury, Vermont, and surrounding region. I could not have asked for a more peaceful and quieter writer's haven.

Introduction

You will come to learn a great deal if you study the insignificant in depth.
(Banzai 2002)

Academics communicate their research to others in their field in a few ways. First, they communicate using the printed word in articles and books, and second, they give talks, verbal presentations. These two types of communication are familiar to most people. Books, newspapers, blogs, and social media posts use writing to inform or persuade. Business keynotes and political speeches show the power of spoken presentation. But a third form of communication is almost unique to academics: poster presentations.

The first international meeting documented to have a poster session was the sixth meeting of the Federation of European Biochemical Societies in 1969, making conference posters at least fifty years old (Rowe 2014). Conference posters may have started in Europe because posters gave people who did not speak the local language a chance to absorb information at their own pace – a distinct advantage considering the range of languages spoken on the continent (Anonymous 2012). Within a few years, the concept of poster sessions had spread to North America:

> The phenomenon was only reported in North America in 1974 at the Biochemistry/Biophysics Meeting in Minneapolis. Not that Americans were slow to embrace the visual, however. The American Chemical Society then introduced poster sessions for the fall national meeting, in Chicago in 1975, a move that resulted in some 41 presentations (take a bow, Divisions of Chemical Education and Inorganic Chemistry). What's more, the session was seen as a "trail blazer." (Anonymous 2012)

But many researchers think conference posters are insignificant. For them, posters are a poor second fiddle to slide presentations. They think posters are mere ephemera: "one and done." Too often, posters are treated as the poor cousin to papers and talks.

But for me, poster sessions are the true beating heart of academic conferences, and I love them. I love the opportunity to talk to people and see what cool new stuff they are doing. I love that posters are often among the first entries in a professional curriculum vitae for many researchers who are early in their careers. I have learned that you can learn a lot about the entire scientific enterprise by paying close attention to poster sessions.

But while I love poster sessions, I have mixed feelings about conference posters. I'm fascinated that they are a presentation form used almost exclusively by academics. But at the same time, I'm frustrated because so many of them are confusing, ugly, or both simultaneously.

Despite this being a uniquely academic field, in typical academic fashion, there is rarely formal training on how to make a great poster. Graduate students are taught writing skills that they need for creating books and journal articles, but because the layout of their published papers is handled by publishers, grad students are unlikely to learn the basics of graphic design, typography, typesetting, and layout. Many grad students are also expected to give slide talks, yet many do that badly, and the task and constraints are different than for a poster. People acquire the skills is through a mix of department folklore, advice from veterans, intuition, and seat-of-the-pants field testing. These are all fine as far as they go, but I would like to provide more than that.

Why should you trust the advice in this book? I expect many readers like to base their decisions on data, but there is not a lot of research on the design of conference posters. Some of the advice here is based on research on other graphic media, like comics, advertising, and web pages, but more advice is distilled from the best practices of many graphic designers. They may not have tested their processes experimentally, but they have developed common practices as a community through hard-won experience. Fortunately, the available research mostly agrees with the experience of designers.

CHAPTER 1

Poster design: the short form

You might have bought this book because you have a conference coming up, you have a poster that you are supposed to present, you need help immediately and are becoming rather desperate. This isn't a great situation to be in, but these things happen.

Here are some guidelines that are extremely robust and will get you to a poster that looks perfectly respectable and that nobody will complain about.

Read the instructions and find out how big the poster boards are. In your graphics software, make your page a few inches shorter in each dimension than the listed maximum.

Divide your paper into three equally sized columns, with 2-inch (50 mm) margins around the sides and between each column (FIGURE 1.1). That is, take the width of your paper, subtract 8 inches (200 mm) for the margins, and divide by three to find your column width. If your poster is 48 inches (1,220 mm), your columns will be 13⅓ inches (340 mm) wide. Yes, it's an awkward number, but computers don't care.

Across the top of your paper, put a horizontal title bar. The size you need will depend on how many words your title is. Type your title in letters more than an inch (25 mm) high. List your name, and the names of other authors, in smaller print beneath that. Do not put anything else in this area.

Consider the remaining space to be roughly divided into six sections: top left, bottom left, top center, bottom center, top right, and bottom right.

Find a high-quality photograph related to your research and put it in the top left or top center section. If you can't find a photograph, find some other kind of image of something distinctive and readily recognizable (not a graph).

3

FIGURE 1.1

A three-column poster layout.

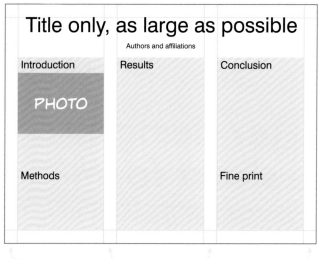

FIGURE 1.2

Final conference poster using a three-column layout.

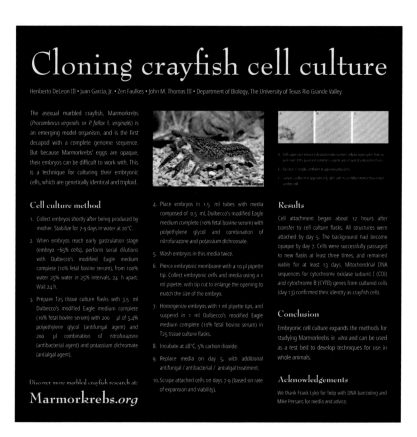

Put an introduction and methods in the left column. Do not use an abstract.

Put your results in the middle column.

Put a conclusion in the top right section. Put references, acknowledgments, and any other fiddly bits in the bottom right section.

Align your text and graphics to the edges of your columns and do not intrude into the margins.

You can hang a poster like this on a poster board of almost any conference anywhere without risk of looking incompetent. You will see many iterations of this style in a typical poster session (FIGURE 1.2). There are still wide degrees of success in pulling off this one simple format. Attention to detail and good choices can elevate this format from competent to stylish.

This is not the only design, nor is it necessarily the best design for your content.

Now that we've done things the quick and dirty way, let's do things right by taking the long way around: a slower, but more polished approach to the conference posters. Let's start from the point of view of someone going to a poster session for the first time.

Chapter recap

- A poster layout of three evenly spaced columns with wide margins might not be exciting but it's hard to screw up.

PART I

For viewers

CHAPTER 2

Attending a poster session

The poster session format

Many people have an idea of what conventions or conferences are like. Maybe they have been to a trade show, a car show, a comic convention, or a business conference. These meetings can have many elements in common with academic or scientific conferences. People expect displays, talks, panels, vendors, and so on.

But poster sessions are almost unique to academia, and they are more common in scientific fields than in others. So many people going to an academic conference for the first time may not quite understand what poster sessions are like.

Normally, poster sessions are held in large rooms or halls: that is, all the posters are in a single physical space. You should not have to run around like a rabbit from room to room looking for posters (unlike oral presentations, which often have simultaneous tracks of programming split across rooms). The posters are mounted on poster boards, with numbers on each board to identify the poster and help people navigate the hallway.

I've heard from some people that they initially expected their audience to show up in a clump, all at once, as though there are guided tours through the conference hall. But a poster session is the "choose your own adventure" part of a scientific conference. Unlike slide talks, which are tightly scheduled, poster presentations are loosely scheduled. Poster sessions normally run for a few hours. In bigger meetings, there can be multiple sessions. Sometimes there is one session a day each for several days. Some conferences have two poster sessions a day. This means you must contend with a time crunch in poster sessions. You cannot just start at poster number one in row number one, talk to the presenter there, then move on to poster number two in row number one. (I've met people who told me they tried this.)

FIGURE 2.1

Overview of
poster session at
2018 American
Geophysical
Union meeting.
(Deep Carbon)

Some conferences split presentation times. For example, people with posters on boards with even numbers present the first half of the session, those on boards with odd poster numbers present in the second half of the session. In theory, this gives poster presenters a chance to see other posters and evens out the load. But in practice, there's no guarantee where a presenter will be at any given time. Some presenters want the full poster experience and will stay for the entire poster session, while other presenters are more *laissez faire*.

This basic format holds true for lots of conferences, but the size of the conference shapes the poster session experience profoundly. At the biggest conferences, when you walk into a convention center, you see row after row of posters almost as far as the eye can see (FIGURE 2.1). When faced with thousands of posters, the sheer size of the space and the volume of material can be daunting. Even experienced conference goers who have attended many small or medium-sized conferences can be shocked by the number of posters at the largest meetings.

To get a valuable experience from a poster session, particularly at the big meetings, you cannot wing it. You need a plan.

Before the meeting

Any time you are going to a conference, set yourself some goals beforehand (Simon 2019). It lowers the intimidation factor of walking into a room that often has a lot of bright, accomplished people, and can prevent you from floundering. If you aren't sure what your goals might be, talk to others, particularly if you have supervisors, who might have some very clear tasks they expect you to accomplish. Besides seeing posters and talks, your goals might include meeting specific researchers in your field, contacting representatives from a funding agency, chatting with the editors of a journal, or talking to vendors about what new equipment and techniques they have.

But there is no reason to show up to a conference without plans to see some posters. Make an itinerary for yourself.

Most conferences compile all the presentation abstracts and provide a searchable database of them online. Your best strategy for organizing your poster-session viewing depends in part on the size of the conference. If it's a very small conference, you might be able to scan all the titles, or at least all the topics. But for larger conferences, skimming through every title may not be feasible or desirable. Larger conferences may group poster presentations into themes or topics. Search both for topics and for the names of authors if you know people in your field whose work you are interested in. The search function in these conference databases is not always as good as in the main internet search engines. Sometimes, they won't ask, "Did you mean …?" if you mistype something. You must supply exact words or phrases.

If the poster session is two hours long, you cannot realistically expect to spend time talking to twenty presenters in detail. That would be six minutes each, taking no account of time for walking, bathroom breaks, or interruptions. A better target is something like four to six posters an hour, which works out at about ten minutes each, plus walking time and interruptions.

Because the number of posters you can see may be limited, rank your list of posters to visit by priority. Go to your "must see" posters first in case the presenter isn't there, so you can try to find the presenter later in the session.

The advance walkthrough

A common way of running poster sessions is that people will hang their posters during the day, often in the morning, and formally present them late afternoon or early evening. This isn't the case for all conferences, particularly the bigger ones, but for conferences that work like this, this is a great opportunity to scan and review posters in a low-key environment. It is less crowded and less noisy.

Walk through the poster hall before the actual poster session begins. The room will most likely be almost empty, so it's easy and quick to navigate. Find the posters you previously identified as interesting. This will help you find them *again* when the real session is going full blast, full of people, busy and noisy, so you don't waste time. You may have to go back to posters a few times to catch the presenters, because they may not be at the poster for one reason or another.

But you can also use your walkthrough to identify posters that you missed in your initial search of titles and abstracts. Conferences are a great opportunity for serendipitous discoveries.

The advance walkthrough is also a chance to start getting a sense of what works in poster design, to inform your own posters in the future. Pay attention to what posters in the hall make you stop and take a second look, and which ones your eyes pass over.

Once you find a poster you have identified as interesting, you can do a quick glance through the poster, and start ruminating on specific questions to ask the presenter during the main session. You may find that some posters are not as interesting as you'd hoped, and you don't need to go back to them during the main presentation session.

Some people will leave business cards, handouts, or other takeaways next to their poster. Going early means that you are more likely to get them before they run out!

If you see someone putting up their poster while you're walking through, be a *mensch* and ask if they need help. Few people are able to reach both corners of a 6- or 8-foot (180–240 cm) poster simultaneously, or see if it's level.

How to be a good poster viewer

Have something you can use to take notes, whether a pen and paper or a smartphone. Having business cards with you can also be a handy networking tool. Carry these by hand or in a small case instead of a large backpack. In the crowded space of a poster session, backpacks make it too easy to bump into other people (Bering 2019).

Most poster presenters will expect that you want a guided tour of the poster. If you have already looked at the poster earlier and have specific questions, tell the presenter so that you can preempt any robotic recitation of their usual presentation. But if you want to hear their prepared spiel, you can just ask an open-ended question. I like to gesture to the poster and ask, "What's to learn here?"

Be attentive. You will be very close to the presenter, so if your eyes glaze over and you start to tune out, the presenter can tell.

Try not to make assumptions about the person presenting, particularly based on a guess of that person's age or career stage. Don't assume the presenter is a student, for example. It's annoying for a presenter to be asked, "Whose lab are you in?" when they are already appointed as faculty in a university. Better questions to ask are, "What's your role in this project?" or "What are you most excited about in this research?" (Fournier 2019). These questions make no assumptions about career stage, they are open-ended to lead to longer conversations, and you will get to hear the things that people are passionate about.

Don't be a creep. A poster session is a professional space, not one to collect phone numbers for hot dates. Some of the creepiest behaviors include standing too close to someone, touching someone, watching someone for a long time before interacting with them, asking to take pictures with someone, asking about their personal details (e.g., their family members), not letting someone exit a conversation, or talking about sex (McAndrew and Koehnke 2016). These behaviors can easily make people feel unsafe, particularly women (see Chapter 24). Think carefully before making any comment about a presenter's appearance, even if you think you are complementing that individual. Comments on appearance are usually irrelevant to the work at hand.

Chapter recap

- The size of a conference makes a huge difference to the poster viewing experience.
- Before attending a conference, set yourself goals and make an itinerary.
- If possible, walk through the poster hall before the session begins.
- Listen actively and behave respectfully.

PART II

For presenters

CHAPTER 3

Why posters?

People who can spread ideas, regardless of what those ideas are, win.
(Godin 2003a)

The overarching goal of a poster is the same as for any other form of communication: to spread your ideas. But posters are a visual medium, maybe more so than any other format that academics use. Academics are a self-selected group of people who are generally much more comfortable than most with reading and writing text. And there has been a tendency in academia to think that everything not only can be represented in text, but should be. It has never been enough to show a picture of a new species of plant or animal, for instance: there must always be a detailed verbal description. The written word usually takes precedence over visuals in academia.

But still, the trend in every medium is towards greater use of visuals. A newspaper from a hundred years ago consisted exclusively of dense columns of text, whereas today's newspapers feature large colored photographs and charts. People spend more time reading newspapers with images than newspapers with text only (Holmberg 2004). Comics rely more on pictures and less on text than they used to (Cohn *et al.* 2017). Many studies of social media consistently show that posts with photos, emoji, or other visuals attract more engagement than those without them (Rogers 2014, Cárcamo Ulloa *et al.* 2015, Pinantoan 2015, Fleerackers 2019).

This means that the skills needed to create posters are going to become ever more valuable in academic careers. The expectations for graphics in slide presentations are increasing. More people are sharing their results on social media, and some academic societies are starting to embrace social media as a platform for full conferences. Academic journals are increasingly asking authors to supply visual abstracts of technical articles (Ibrahim *et al.* 2017, Gloviczki and Lawrence 2018, Ibrahim 2018).

Posters versus papers

A major part of the training and professional expectations for academics is that they learn how to write papers in the format expected for their research field. The training academics get in writing papers is so powerful that many people default to duplicating the structure of a technical paper for a scientific journal, even when there is no need to. Academics are so accustomed to writing papers that they ape the format for posters slavishly. As the saying goes, "When the only tool you have is a hammer, every problem looks like a nail." And academics are very well trained to use their metaphorical hammers.

A poster is not a paper. Most academic journals have extensive, rigid instructions. Posters rarely do. You do not have to duplicate everything you would do in a technical paper on a poster. An abstract on a poster serves no clear purpose. Nobody expects the methods to be laid out in so much detail that they could replicate the entire experiment from the poster. Nobody expects a poster to have the same exhaustive reference list a paper would have.

Academics are often expected to be able to describe everything in words. This made sense when images were difficult to create and expensive in printed journal articles. But relying on words alone is a bad look on a poster. Creating a poster calls for a different way of approaching the material. **Posters are a visual medium.** You must think graphically, not verbally.

Posters versus talks

Of the three ways academics communicate (papers, talks, and posters), only talks compete directly with posters at conferences. Presenters usually have a choice of whether they want to give a talk or a poster, and audience members usually have their choice of whether to attend a talk or a poster session. Most people prefer talks over posters, both as a presenter and as an audience member. I've seen this result in several polls, including some I've run myself. Usually, about 66–75% of people say they prefer giving talks at conferences over posters.

Many see posters as a second-rate form of conference presentation. But some of the reasons that people give for preferring talks over posters are not very good reasons.

One objection is that posters take more work than talks. I have no hard data on this, but I have seen very effective posters that were completed in less than an hour (according to their creators). But this complaint may be more about the timing of the work than the amount. The hidden message is usually, "I can work on slides for my talk on the plane trip to the conference, and in my hotel room." Sometimes, slides get tweaked in the coffee break before the talk, which you can't do with a poster. Once a poster is printed, you can't make changes. But this is a point in favor of posters. As much as procrastination-prone academics might love to believe that a plane trip is "productive time" because they were working on a presentation, knocking out a slide deck on the plane is not likely to lead to a high-quality talk. If you want to make a plane trip productive, read some recent papers from a journal in your field.

People also say that they can reuse slides from one talk to another. But posters often tend to be modular in design, and text and graphs can be imported from one to another without too much effort.

There is a greater possibility of being "shut out" in a poster session than a talk. In most conferences, there will almost always be someone in the room during a talk, but nobody can guarantee someone will talk to a poster presenter. But good poster design and strategies can help attract an audience.

Lots of people choose the format of a presentation based on the progress of the presenter. For whatever reason, people tend to think of a poster as either "less stressful" or "less important" than slide talks, so senior researchers tend to encourage early-career researchers to use a poster session as a first conference presentation (McNutt 2015).

A better approach is to look at the content of the project and the venue rather than to assume that all new projects should be posters. There are times when a talk is more appropriate than a poster, and vice versa. The two formats serve different purposes. Here are a few factors to consider in whether your project is better as an oral presentation or a poster presentation.

Does your project have a single main point or result? Many scientific posters suffer because people have done many experiments, and they think they need to show every single one of them. A dozen experiments may each need their own graph to explain the results. Many graphs from many experiments lends itself more to a talk, when you can show one graph at a time. But other projects may have results that can be shown in a single image. That image may be very dense and contain a lot of infor-mation, like a diagram showing the relationships between many different species, or a map showing the distribution of something. A poster gives people the time to drill down to find the details in the image in a way that would be very difficult if it were presented during a talk.

How far along is the project? A poster is often a better option when a project is partially completed. A new project will often have less data,

FIGURE 3.1

Project results
should determine
presentation form.

MANY SIMPLE GRAPHS = BETTER AS A TALK

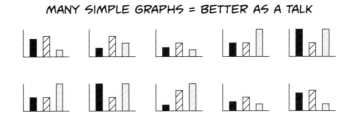

FEW COMPLEX GRAPHS = BETTER AS A POSTER

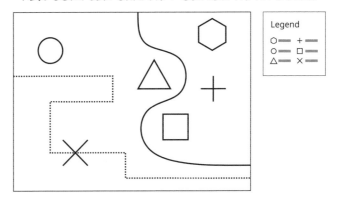

which makes a focused poster easier to make. You think you have a good story, but you're not quite sure if there are missing pieces. You have a puzzle, and you're not sure quite sure if the next step you want to take is the logical one. This is where connecting with people for one-to-one discussions is valuable.

Another consideration for giving a talk or poster is the size of the meeting. If the meeting is very small, there is probably only one track of programming. A talk may be better because you are potentially able to reach a higher percentage of the audience. You would probably get a chance to see everyone you need to see regardless of whether you give a talk or a poster. At medium-sized meetings, there are often concurrent talks, but a few concurrent sessions might still mean a potentially larger audience than a poster session. But a poster becomes more attractive as the conference gets bigger, because you're more likely to hit your target audience. At the few very large meetings where attendance is in the thousands or tens of thousands, posters are a more effective way of connecting with the people most interested in your work. There may be a four-hour window for a poster presentation compared to only a fifteen-minute window for a talk. And given how many things are going on when a conference has tens of thousands of people, you stand a better chance of connecting with people the longer your presentation runs.

What is your own comfort level with speaking in front of strangers? "Speaking in public" always ranks highly in lists of things that people fear. Some people may be so terrified of speaking in public that they will always opt for a poster. A poster presentation may be a much lower stress form of presentation. But some people may be more comfortable with an oral presentation, where they are somewhat shielded from potentially scary or awkward one-on-one interactions with an audience member. Someone who has anxiety may be more stressed by the prospect of talking directly to someone they see as "famous" than knowing in an abstract way that someone "famous" might be in the audience.

Is your goal to promote or discuss? Oral presentations are a great opportunity for evangelism. History has proved that great oral presentations (usually better known as "speeches") can change the world. If your goal is to excite or rally, an oral presentation is probably the way to go. But to promote and inspire is not always the goal. Sometimes the goal is to discuss and connect. This is where a poster presentation can benefit the presenter, because a poster presentation lets your audience **talk to you**. When you present a poster, you are physically close to your audience, and you are typically talking to only a handful of people at a time. This is much more conducive to audience members asking questions and volunteering ideas.

For example, graduate student Mya Roberson (2018) describes this affecting conversation in front of her poster:

> As I stood next to the poster detailing my work down to the level of molecular characteristics of breast cancer, researchers from all over the country stopped to question me. I took copious notes, eager to capture a fraction of the ideas and inspiration in the room. I even met some of my academic idols. I felt like a real scientist, engaged in the process of scientific inquiry.
>
> Later in the session, a black woman with silver hair and no institutional affiliation listed on her name tag approached me. Before I could begin my standard elevator pitch, she said she'd prefer to read my entire poster first. I studied her face as she read. A few minutes passed. The concentration in her face gradually shifted to raw emotion.
>
> After she finished reading, she introduced herself as a survivor advocate, someone who is not a scientist but interacts with them to give the patient and survivor perspective. She shared her medical history – her diagnosis of severe endometriosis in her early 30s, the removal of her ovaries and uterus in an attempt to allay the symptoms, and her more recent diagnosis of breast cancer, which led to her advocacy in the cancer research community.

Then, she reached out and touched my poster and said, "I see myself in this research. This was a study meant for women like me."

Her personal revelations stood in stark contrast to my previous interactions with fellow scientists, which were mechanical and formulaic. This was different. It was two black women talking about our resilience. […]

Now, when I present my analyses of the binary numbers representing women who gave part of themselves for the advancement of cancer research, I include her story. To me, her interaction with the humanity of science is just as important as the output of my statistical models.[1]

Slide presentations are monologues (sometimes brilliant ones); poster presentations are conversations. Poster sessions are places where you might see warm greetings, casual conversations, and jokes in addition to the exchange of technical information. Nobody ever opens their laptop and starts working on their grant proposal while you're talking to them about your poster. That's why posters are better than talks.

Chapter recap

- The skill set required to make posters is increasingly more valuable and can be applied to many academic activities besides posters.
- Posters should not be treated as papers on one large piece of paper.
- The choice of giving a talk or a poster depends on the type of project, the size of the conference, the presenter's comfort level in different situations, and the goals of presenting the project.
- Posters facilitate one-on-one conversations.

1   From *The New York Times*. © 2018 The New York Times Company. All rights reserved. Used under license.

CHAPTER 4

Design thinking

For objects like conference posters, people often think that "design" roughly means "artistic," "decorative," or "creative." There are two problems with this. First, it reflects the bias of some academics who disdain beautiful objects (Antonelli 2007) because they equate them with fluff. "Some assume that an aesthetically appealing presentation signals at best a lack of priorities, and at worst a lack of rigour" (Krause 2019). Second, art and creativity isn't design (Rutledge 2008, 2009). Art is unconstrained and need not answer to anything but the whims of its maker. Design is always constrained and must answer to the person using the item.

Design is a *process* for making items. Almost everything that humans interact with is designed. This book was designed. Your clothing was designed (and you look mighty snappy in it, I must say). Your kitchen cutlery was designed. Design at its best incorporates an aesthetic sensibility, but even the most purely utilitarian tool is also designed.

Defining design

Design can be summed up in a phrase: decisions with empathy.

Let's explore that phrase, in reverse order.

Empathy is about thinking about who will be using what you are making. You shouldn't make something a certain way just because it is convenient for you; you should make it that way because it is convenient for the user. For a poster, you can think of this as your reader or your audience or your fellow conference attendee. You probably have an advantage here, because a lot of people going to the conference will have similar interests to you. The bigger the conference, the less likely this is to be true. If you have been to a conference before, you know that it can be exciting, interesting, and exhausting – all at once.

The first part of my definition is about making deliberate choices. Everything in a designed object should be the result of a considered choice. Why this size, this shape, or this color? And there should be clear answers to those questions, because most humans perceive things in roughly predictable ways that are described by psychology (Rutledge 2009). But even if you haven't made a deep study of, say, the psychological concept of gestalt to justify your choices, you should at least make your own choices, not leave them up to anyone else.

Software applications contribute to poor design because they bypass decisions. Because most posters are made with computer software, when I ask people why some aspect of their poster looks the way it is, the answer is often, "That was the system's default setting." Because so many people can't be bothered to change default settings, it's easy to end up with a poster that looks like all the others.

Templates also contribute to poor design for the same reason: people stop making decisions. Lots of people want templates, and I understand why. People want to save time and decision making is hard, and templates save time by removing many decisions. But using templates usually causes more problems than the amount of time saved, for several reasons.

First, many templates are surprisingly poor. I have seen templates for posters from universities that are much more concerned with promoting the university's name than the actual content of the poster. Many templates encourage the "wall-of-text" approach to a poster.

Second, many templates don't fit the needs of a project. Online templates in landscape format aren't much good if the conference is requiring portrait format.

FIGURE 4.1

Billboard-style poster template. (Mike Morrison)

Third, if you don't understand why a template is the way it is, you risk screwing it up any time you change something. For example, one billboard-style template consisted of three parts (FIGURE 4.1): a structured abstract on the left, a big take-home message in the middle, and fiddly details on the right (an appendix or supplemental material) (Morrison 2019). Some people who used the template kept the big central message but split their essential material between the left and right columns. The right-hand side was no longer the optional "fine print." Because the content was separated as far apart as possible, it made the story harder to read.

An example of the importance of graphic design and the risks of following templates is provided by Edward Tufte, who made a compelling argument that in one case, bad graphic design may have contributed to deaths (Tufte 2003, 2006). FIGURE 4.2 is a slide that was used in analyzing the risks posed during a mission of Space Shuttle *Columbia* (Parker *et al.* 2003). The spaceship disintegrated on reentry and the entire crew died.

This slide is clearly based on a PowerPoint template. It boasts six levels of hierarchy: a title, four levels of bullets, and parentheses inside text.

This slide's title suggests "everything is fine." But the design of the slide obscures a much less confident conclusion. The second-last line admits, "Flight condition is significantly outside of test database," which says the conditions have wandered into uncharted territory. In retrospect, this should have been the most important thing, not near the bottom. Among many other faults (e.g., "Initial penetration to described by

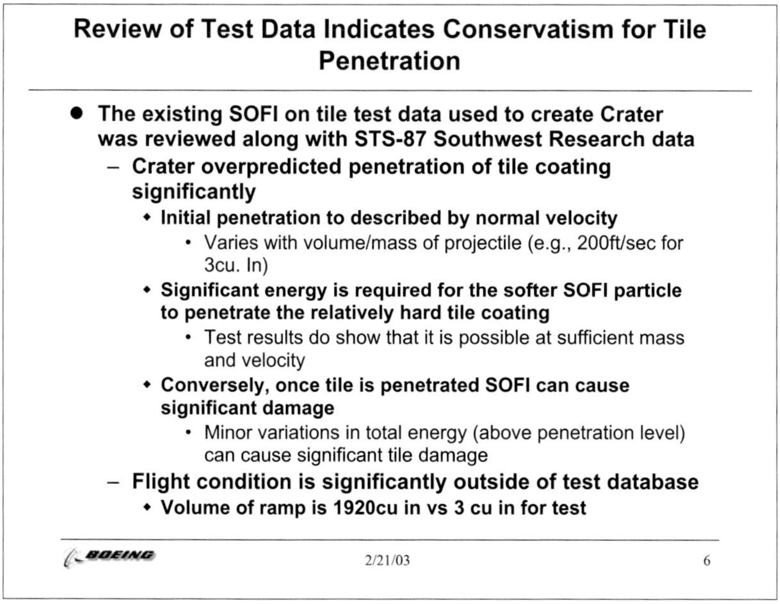

FIGURE 4.2

PowerPoint slide used in NASA briefings about Space Shuttle Columbia.

normal velocity," which is non-grammatical at best and nonsensical at worst), units are shown inconsistently, which indicates a lack of attention to detail. While NASA clearly had many cultural problems throughout the organization that contributed to the Shuttle disaster, the muddled slide design represents unclear thinking.

These are examples of why core skills in graphic design include "analysis" and "research." Other core skills are layout, typography, composition, color theory, and software skills (Dabner *et al.* 2010). These skills contribute to larger design literacy: noticing the design of any sort of human object, even boring and mundane ones; analyzing designs; and expressing yourself in a visual way (Kolko 2018).

Design brief

Design works best when you have clear ideas about what you are trying to accomplish. When professional designers take on a task, they get a design brief from their clients that specifies the goals, the audience, and the desired outcome. Many problems stem from the poster creator's design brief being misaligned with expectations of what audiences want. Expectations of what poster viewers want seem to be based on clichéd stereotypes of academics that people carry around in their heads.

When I ask people what the goal of a conference poster is, they usually say something like, "To give information." Consequently, they think that the poster with the most information on it wins. They cram the poster full of as much detail as possible.

Posters do convey information, but that's not the most important thing a poster does. The most important thing a poster does is to start conversations – especially conversations with strangers. Conversation is a deep part of human society. We learn to speak before we learn to write, and children pick up sophisticated grammatical rules with no explicit instruction in those rules. "If you diagrammed a conversation, it is amazingly complex. Conversation is more complex than any writing, yet it is more likely to lead to understanding" (Wurman 2001).

The poster that starts the most conversations wins, not the poster that has the most information.

But different people have different opinions about what they think the goal of a poster session is and what the goal of a single poster should be. For example, the template in FIGURE 4.1 was designed for a viewer to be able to walk through a hallway and read a lot of conclusions in a short amount of time. This is good design because it takes what viewers might want into account. But whether this format is successful depends on how many viewers *want* to skim through large swaths of poster boards to absorb as much diverse information as possible. Many viewers don't want to do this. Instead, they want to take a much more focused approach. They want more detail on the posters they consider most relevant to them so they can have a "deep dive" on the topic with the presenter.

Unfortunately, because people want different things from a conference, a poster session, and a single poster, no single poster design can be optimal for absolutely every viewer. But careful decision making can help broaden that usefulness for more viewers, so that even if a poster design isn't perfect for one viewer, it's close enough to perfect that said viewer might forgive its flaws.

Talking to strangers

Any time you give a presentation, there are going to be people who are already interested in what you have to say (an inner group). These are people who either know the topic professionally or may know you personally. Because of their social connection with you, they are liable to come and talk to you regardless of where your poster sits on the information continuum. While these people are important, if the goal is to spread your ideas, they are not the only people you should talk to.

There is a much larger selection of people who are – at least initially – not interested in what you have to say (an outer group) (Olson 2019a). Most of these people are going to be strangers to you. These are new people who you can spread your ideas to.

The point of academic conferences is partly to get talking to people you have never met before, but this is something that most of us hate doing. As Nina Simon (2008a) said, "We spend most of our time studiously ignoring strangers, and it takes extraordinary situations to overcome those cultural mores and fears." Teenagers said they would go to great lengths to avoid talking to strangers, and that they wanted other people to start the conversation (Simon 2010a).

One reason people will talk to strangers is when a stranger has something that is visible and worth talking about (Godin 2003b). For Nina Simon, her dog was a remarkable object. Dogs are worth talking about! "When I walk around town with my dog, lots of people talk to me, or, more precisely, talk through the dog to me … It's much less threatening to engage someone by approaching and interacting with her dog" (Simon 2010b). Even a crappy polyester vest that identified Simon as staff could be remarkable, because the vest made her someone who was safe to talk to (Simon 2009). To put it another way, both Simon's dog and a conference poster are *social objects*: things that people want to share with each other (Cetina 1997, Engeström 2005, Simon 2008b).

People are more comfortable talking about some third object than they are talking either about themselves, or about the person they are supposed to talk to. After all, talking about yourself, or the person you're talking to, is not an easy thing to do. It requires some trust.

What starts conversations? Above all else, you need something recognizable if you want to start a conversation. If there is nothing a novice can comprehend at a glance, you're not going to have conversations. But you have an even better chance of starting conversations if that entry point is remarkable. That remarkable thing can be an amazing fact. It can be a tantalizing question. Or it just might be something unusual, edgy, or unexpected.

Competing for attention

We live in an attention economy. When there is too much stuff and not enough time to absorb it all, one of the most valuable things you can give is not money, but attention (Visocky O'Grady and Visocky O'Grady 2008, Olson 2019a). People are continually making decisions about what to pay attention to, and they filter based on what they deem relevant to them. "In today's short-attention-span world – you bore, they snore, you confuse, you lose" (Olson 2019a). Salespeople have competed for attention (and cold hard cash) for a long time, and they have a model for engagement. This model is sometimes called the "sales funnel" or "awareness, interest, decision, action" (AIDA), shown in FIGURE 4.3.

The AIDA funnel recognizes that there is a predictable series before people act. But it emphasizes that potential customers (or poster viewers) are lost every step of the way. Common obstacles are shown in FIGURE 4.4.

The physical environment and mind-set of the audience both challenge designers. Many conference poster sessions are crowded, noisy, badly

FIGURE 4.3

The AIDA sales funnel.

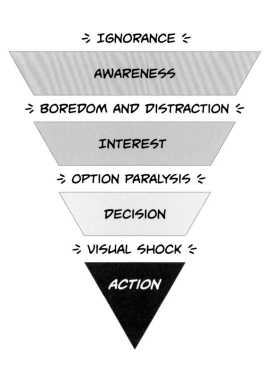

⇉ IGNORANCE ⇇

AWARENESS

⇉ BOREDOM AND DISTRACTION ⇇

INTEREST

⇉ OPTION PARALYSIS ⇇

DECISION

⇉ VISUAL SHOCK ⇇

ACTION

FIGURE 4.4

Obstacles to
that break the
AIDA funnel.

lit, and have a bar to one side serving alcohol. Conference attendees are a mix of people. Some are new to these events and feeling confused, excited, or both. Some are veterans who are bored and grumpy and feel like they should be back in their hotel room working on a grant proposal. And both newcomers and veterans may have traveled long distances and be tired or jet-lagged. "Look, read, listen, think: the audience of your presentation doesn't want to do more than 2/4 of these at any time. Nor should they" (Krzywinski 2019).

A lot of the people at academic conferences are relatively new. They are graduate students and postdoctoral researchers. For many, walking into an academic conference is as potentially confronting as walking into something like a place of worship for another religion: unfamiliar and a little discomforting (Simon 2016). That means they are less familiar with navigating conferences, as well as with some of the concepts and ideas presented, and conferences are just generally more work for them because of their newness.

Most posters probably start with roughly same level of **awareness** among conference goers, since all posters are listed in the conference program book and are visible to anyone walking through the poster session. You might gain a little extra awareness if your abstract is the first in the program book, or by tweeting about your poster. You lose a little audience awareness if your poster is hung in a remote corner of the hallway the session is in.

Some of obstacles are largely out of your control. But the design of your poster is something you can control.

What determines **interest** for a poster viewer is probably going to be that person's assessment of relevance. What people deem to be "relevant" is more likely to be shared on social media, for instance (Botha and Reyneke 2013). What someone considers "relevant" will vary from person to person. But there are two factors that make information relevant: it helps you generate new conclusions that matter to you, and it's easy for you to get the information (Wilson and Sperber 2004, Simon 2016). That last part is crucial, because it is a reminder that "relevance" isn't a fixed, unchanging property of information. It bears repeating:

The amount of effort that people need to put in to getting information changes its relevance for them.

If it's too hard, people will give up and brand that information as irrelevant. Even when explicitly searching for information that they want (i.e., that they deem relevant to them on the face of it), people follow the Principle of Least Effort (Mann 1993). Not everyone will read all the poster abstracts in advance if it takes less effort to walk through the hallways and browse. It is critical that you make it as easy as possible for people to get to the information that will help them have those insights.

For example, reading is a complex, time-consuming task. We academics tend to forget this, because we are self-selected group of people who liked reading and writing so much that we aspire to make a living from it, and sometimes succeed in doing so. But nobody is born knowing how to read. Few children can pick up reading by observation in the same sorts of ways that children learn spoken languages. It usually takes years of instruction and purposeful, directed effort to learn how to read. It takes even more years to learn how to read complex technical texts, like the sort found on most conference posters. Too many posters create "walls of text" that people are expected to read (Greenfieldboyce 2019, Morrison 2019). This can increase the effort needed so much that it dooms a poster to irrelevance.

Further, complex diagrams can overwhelm viewers. They do not know where to start or how to analyze the diagram. This was called "map shock" by researchers studying "node and link maps" (flowcharts) (Blankenship and Dansereau 2000), but the reaction happens when people are faced with any complex data visualization. A more general term might be "visual shock" (Visocky O'Grady and Visocky O'Grady 2008). Visual shock makes people lose motivation to understand the diagram such that they do not process it fully. General Stanley A. McChrystal was in visual shock when presented with a complex "bowl of spaghetti" flowchart of American military strategy in Afghanistan, which led him to say, "When we understand that slide, we'll have won the war" (Bumiller 2010). His joke showed his lack of interest in analyzing the image to understand it.

If you make your poster hard to understand, people will judge your poster as irrelevant, and decide not to talk to you. Good design can convince a viewer that there is some information that they can extract easily. For example, the poster in FIGURE 4.5 is concise and easy to parse, which helps the viewer make a **decision** and take the **action** of talking to you.

People share unexpected or surprising content on social media (Allocca 2011, Berger and Milkman 2012, 2013), so non-standard presentations

FIGURE 4.5

A concise poster.
(James C. O'Hanlon)

make posters more likely to be shared. Examples of conference posters that have received attention because they are unexpected in design or content include:

- A poster with blacked-out content and a note explaining that the presenter could not attend because they had been denied a visa (Dayas 2018, Rodriguez 2018).

- A poster with "plunging" in the title with an illustration of a bra (Wilson 2015).

- A poster that said, "Ask me about my cat!" with a picture of said cat. This was followed by, "and while you're here, have a look at our study" (Knowles 2019).

Catching attention is not guarantee that reaction to a poster will be positive, however.

Emotional content is most likely to be shared (Ogilvy 1963, Berger and Milkman 2012, 2013, Botha and Reyneke 2013). But not all emotions are equally motivating. Something that causes high arousal (e.g., agitation rather than relaxation) and positive emotions are more likely to be shared (Berger and Milkman 2012, 2013). Something that makes people sad is less likely to be shared (Berger and Milkman 2012, 2013). "Awe" might be the best emotion a poster maker might aim for.

One way to create that emotional response is to make your poster beautiful. You need to entice people, make them stop in their tracks, and invite them in. (This will be a recurring theme in this book.) This is something some scientists sniff at. Paola Antonelli (2007) said, "Many scientists tend not to use anything beautiful in their presentations, otherwise, they're afraid of being considered dumb blondes." Some academics think that making something beautiful means losing the integrity of the information (Meeks 2018). Some think there is no need for any aesthetic considerations, believing "the data should speak for themselves." But data never do. Data always need advocates.

Part of the problem may be that scientists think that making something beautiful means decorating it with irrelevant frippery. People might imagine turning a serious scientific study into something that looks more like a wedding cake. But a plain object can still be beautiful if it is well designed. Shaker furniture is undecorated but still valued for its construction and elegance.

Admittedly, people might think that aesthetics corrupts because so many "artsy" posters are ham-fisted, amateurish, and horrible to look at from a design point of view. But don't judge the field by its worst practitioners. Scientists never get formally trained in this. It's much better to study and embrace what professional graphic artists can teach us, and to strive for graphic excellence in all presentations. Making ugliness a virtue is the wrong way to go.

Creating a poster with the express purpose of breaking the mold or provoking an emotional reaction carries risks. You may create something that people hate, and "anger" is one of the most powerful incentives to share something (Berger and Milkman 2012, 2013). Most people want to avoid controversy rather than court it. It's also worth remembering that the goal is not simply to get eyeballs on the poster for any reason, but to increase your conversations about the ideas on the poster.

Conveying information

If you get someone's attention, how long will you have it and what do you want to do with it? You want your ideas to spread, and this will mean conveying some information. How much information? We can think about the two extremes. One the one hand, there is very dense information, the "journal article on a wall" format shown in FIGURE 4.6.

This is an exaggeration – but not as big an exaggeration as you might think. I have seen pages of manuscript text tacked to a poster board. Luckily, this was always rare and has become even rarer.

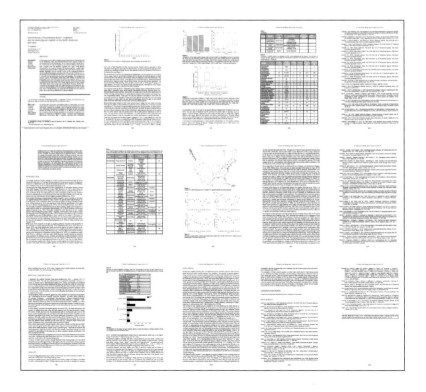

FIGURE 4.6

A mock-up of a poster literally made from a journal article stuck on a poster board.

FIGURE 4.7

A poster design that invites discussion but has no details.

The other end of the continuum is shown in FIGURE 4.7, where there is almost no information. It is friendly but it is not informative. I don't know if anyone has done it, but at least one researcher wanted to make posters like this (Tennant 2014).

It seems likely the "wall-of-text" design is unlikely to meet the needs of an "outer circle" viewer because it requires too much effort. The "talk to me" design is also unlikely to meet the needs of a viewer because there isn't enough detail to motivate a viewer to talk to a presenter.

Because design requires practicing empathy, one of the key questions is, "How much information does a poster viewer want?"

We can get a handle on how much information people want by asking people how long they want to spend talking to a poster presenter (FIGURE 4.8; Faulkes 2019).

FIGURE 4.8

Poll results to "How long do you want to spend talking to a presenter at a conference poster?"

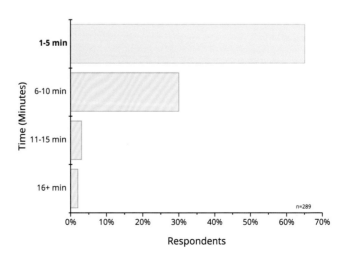

About two-thirds of poster viewers (65%) want to spend no more than five minutes viewing a poster and talking to a presenter. Just under a third (30%) want to spend five to ten minutes talking to a poster presenter. Only about one person in twenty (5%) wants to spend more than ten minutes (Faulkes 2019). People are willing to spend more time if necessary (Crowe 2019), but this is a good benchmark for how much material to include. If you can't explain it in five minutes, you probably have too much stuff.

It's worth noting that the typical length of time that viewers want to spend – five minutes – is less than half the time allotted to typical slide presentations at conferences, which usually run 12 or 15 minutes. You should not plan on presenting the same amount of content on a poster as in an oral presentation.

Despite this short time frame, academics will often say they want a lot of the supporting details on a poster. Academics are trained to be critical (and are sometimes a little obsessed with proving how smart they are). They want to be able to see the data supporting conclusions so they can pick it apart to see if you made a mistake. They want to be able to decide whether they personally would arrive at the same conclusions.

But people will often say they want one thing when asked, but do something rather different in the moment. The description of a thorough and skeptical poster viewer probably represents an *idealized* poster viewer. But the *realized* poster viewer – the one who has traveled, sat through other talks, is looking for a drink, and has lots of other posters they want to see – might not be as concerned with the details as they claim from the comfort of their offices. The *realized* poster viewer might be a happier if you can pare back the details and give them with a clear take-home message. While some complex representations may be familiar to experienced people in the field, conferences usually have newcomers trying to enter the field. If you give these people something more accessible, they will feel like one of today's lucky ten thousand (FIGURE 4.9; Munroe 2012).

When people think of posters – not conference posters, but any other sort of poster – they usually think of them as works of art. Movie posters or concert posters are remembered as art, not as advertising. Wartime posters like Lord Kitchener or Rosie the Riveter are remembered for their art, not as propaganda. Those posters are all brilliant works of art. Some academics think their conference posters should also be works of art. But while good conference posters may be artistic, they are not art in the way that people usually think of art: something generated by creativity and personal expression. Plus, the classic posters mentioned above work so well partly because they are simple. They convey one statement. Research posters are rarely that simple.

FIGURE 4.9

"Ten thousand,"
from the xkcd
webcomic.
(Randall Munroe,
https://xkcd.
com/1053)

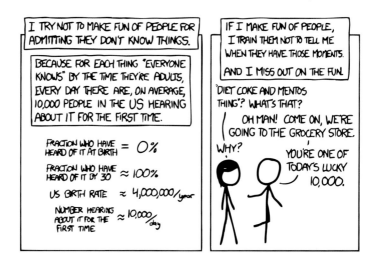

Conference posters are documents. They are large documents. They are illustrated documents. They are documents that benefit from concision and simplicity. But they are documents nevertheless. They are just as much documents as technical reports, executive summaries, personal statements, or journal manuscripts.

Thinking of conference posters as documents clarifies the task at hand. We associate "documents" with conveying complex information. Craft and thought are necessary for the job at hand, more than creativity. The craft is graphic design. There is a great body of theory, principles, and thought associated with layout and typography. I'm sure that most researchers have never studied these at all, based on the apparent disregard for even a simple grid. I see poster after poster where you're lucky to see two objects out of twenty on the paper align with each other.

And here we arrive at the point where most people probably started: that a poster is a way to convey information. But this is the last point in this list and not the first because, simply put, thinking about all the information you want to convey on your poster will do you no good if you're not interested in talking to people and you can't get their attention.

Accessibility

That posters are a visual medium has pros and cons. On the plus side, so much graphic design is rooted in common human physiology and psychology that "it breaches most boundaries. Often, you do not need to speak the language to understand. Good design translates into most languages and cultures" (Landa *et al.* 2007).

But visual media can be frustrating for people with visual impairments. It may not be possible to make a poster that is completely accessible to everyone with any visual problem, but a few good choices will stop you from accidentally shutting out your potential audience.

Presbyopia is common in people who are middle-aged and older. They have problems reading fine print and need reading glasses. But there are several solutions that are relatively simple and help everyone. First, make sure everything on your poster is big. Second, pick typefaces that are uniform in thickness. Third, make sure words are printed against high-contrast backgrounds.

Dyslexia hinders people's ability to read text. Dyslexia affects different people in different ways, so there is no simple checklist that will optimize a poster for all dyslexic readers (Collinge 2017). But there seems to be general agreement that careful attention to spacing and using sans serif fonts will help many dyslexic readers (Collinge 2017, Morris 2017).

Color blindness affects more men than women and comes in several forms. Designing for people who have difficulty distinguishing colors will be discussed in more detail in Chapter 10. Techniques to make posters more accessible for people with other visual impairments, like low-resolution vision, will be discussed in Chapters 12 and 17.

Let anarchy reign!

Academics get used to receiving obsessively detailed instructions when they are writing. Journals have exacting instructions for authors that we are expected to follow before submitting a manuscript. Grant proposals are worse. These instructions and standards often go on for pages, detailing expected sections of text, typeface, point size, margins, and – everyone's favorite – how references should be formatted.

Compared to those, poster sessions are like the Wild West. They are nearly lawless environments. Many conferences do have guidelines and recommendations for their conference posters, but there's usually only one hard and fast rule, imposed by geometry more than conference organizers: posters have to fit into the designated space on the board.

If you catch yourself asking, "But can I do …?" when you're thinking about designing or presenting a poster, remember that **you are free**. You are free to ignore recommendations. You are free to change the title or the planned content of your poster if it makes sense to do so. In most cases, the only thing guiding you is your own sense of right and wrong – in the design sense, that is.

Occasionally, when I point out some aspect of a poster design that I think could be improved – usually an abstract or a logo – people justify it by saying, "It was required." But the reality is that they are requirements in name only. Nobody checks the poster before it goes up. And nobody will come by with a clipboard to look at your poster and check that all the required elements are there. There are no poster police.

There are, however, sometimes poster *judges*. We'll talk about how to win a judged poster competition in Chapter 5.

Chapter recap

- Design requires empathy.
- Your design brief: You are competing for the attention of viewers, who may have visual problems. The poster that starts the most conversations wins.
- There are no poster police. You can generally do what you want.

Early preparation

Most posters fail early. People don't put enough thought into either the content or the design of their posters at the early stages of poster making. If you don't know what the main point of your work was, you are going to fail no matter how good the design of your poster is. If you pick a nonsensical layout for your poster or your poster won't fit in the allotted space, you are going to fail no matter how good the content is.

Discuss with colleagues

Many posters result from collaborative research programs. It's common for someone early in their career to make a poster that must please many collaborators who may be more senior and have significant power over the poster maker. Make sure that your collaborators are all on the same page and agree on key points.

First, your colleagues may have very different ideas about what the most important point or finding is. By having lots of discussions, you can avoid having to redo the entire poster because one team member says, "I don't think that's the right way to go."

Second, many researchers will not agree with many of the recommendations and advice in this book. They might think graphic design is completely irrelevant, or they may not have thought deeply about graphic design. There can be inertia ("We've always done it this way"). Simply put, your colleagues may want you to make a bad-looking poster. If you are a student, you may feel pressured by your supervisor to make a poster in a way you don't agree with.

Third and most important, make sure all collaborators agree about who should be included on the author list, and what order the authors should be listed in. Conference presentations are relatively small stakes in professional career building next to papers, but poster presentations can be the first lines of a professional curriculum vitae for many students. Posters can create a track record that improves the odds when applying for scholarships, new programs and other professional opportunities.

Many researchers are afraid of discussing authorship credit (Smith *et al.* 2019). But if there is no discussion, arguments over authorships are a completely predictable result. As hard as it may be, asking direct questions like, "How is authorship credit for conference presentations determined in this lab?" are necessary conversations to have. You do not want to be in a situation where you get less credit than you think you deserve. People remember their disappointment about not getting credit and can be bitter about it for *years*.

Authorship practices vary from field to field and even from lab to lab. In particle physics, authorship is usually alphabetical. But in many fields, first authorships matter most, even on posters. In life sciences, the first author is presumed to have done most of the work and the last author is usually the oldest person who supervised the work. The ordering of the middle authors is usually from largest to smallest contribution, but nobody cares enough to check.

While there are many subtleties to author lists (which is why you need to talk about them), there are a few good practices that most people agree on. First, people who did the work presented on the poster should generally be included as authors. Second, everyone who is listed on the poster should know they are listed on the poster and agree to their name being there. Third, people who did none of the work presented on the poster should not be included – a practice sometimes called "honorary" or "gift" authorship.

Seek professional help

Decades ago, a technical document or presentation used to represent the work of many people (Duarte 2009a). A complex report from a business or government might have been generated by an author, a statistician, an editor, a photographer, an illustrator, a typesetter, a proofreader, a printer, and so on. A label in a museum or art gallery might require a writer, editor, and designer (Serrell 1982). Part of this specialization was because the professional tools that people used were expensive and not everyone had access to them. Only someone deeply interested in photography would invest in a darkroom, for example.

But the rise of personal computing put professional-level tools in the hands of almost everyone. A typical smartphone has a camera as good as or better than the cameras that professional photographers used to use. Digital photography eliminated the need for darkrooms. Graphics software meant people didn't need to go to art shops for tools to draw smooth curves. It became possible for everyone to at least try tasks they could not have done in the past because they didn't have the equipment.

Because they have access to those tools, many academics expect that they should be able create a flawless technical document or presentation alone. But those powerful professional tools don't come with skills, and learning the skills takes time. Therefore, consider getting help from people who have skills that you do not. Look further than your own immediate circle of colleagues and department. Your colleagues may be self-taught and guided by "We've always done it this way" rather than being informed by targeted training and engaged practice in a professional community. It can pay to look around more widely for experienced people with a professional skill set who can help with your poster. If you are in a large institution, or on a university campus, you may have a lot of resources available to you if you ask.

Public relations departments in your university may have a style guide. Style guides specify university colors (see Chapter 10) and guidance on preferred name and phrases. For example, is the official name of your institution Erewhon University, University of Erewhon, or *The* University of Erewhon?

One of the most useful things you can have on a poster is a photograph of whatever the poster is about. Photographers are a typical part of university staff. Universities are always needing photos of events for news and promotion. Contacting a staff photographer might be a great way to get a compelling visual for your poster. Photographers are often happy to do this, because they are looking for shots of faculty and students that can be used for university promotional materials.

Many universities have dedicated graphic designers. Who do you think makes all those campus flyers and promotional material? These offices are sometimes in university relations, news and public relations, or some other place, so they may be hard to find. But most universities are going to have staff tucked away somewhere who are busy creating the look of documents the university puts out. And universities put out a lot of documents. The downside of working with institutional graphic designers is that they may be more concerned with institutional branding than developing a look that is right for your poster.

Illustrators may not be common as part of university staff, but faculty in the arts may be able to assist you either by providing illustrations or by recommending someone with the right skill set.

Some bigger campuses have their own campus press, with people who are experienced in technical printing and design. There are often large-format printers available so that you do not need to send files to printers in another time zone and have posters shipped back to you by courier.

The above focuses on professionals who may already be on your campus and willing to help you as part of their job description, but another option is to work with professionals who have no affiliation with your institution. Many artists, designers, and photographers work digitally and remotely, so your options are not restricted to people working locally. And they are often actively looking for commissions. Look around for people who specialize in the kind of work you are doing. Some people specialize in accurate renditions of animal anatomy, while others might specialize in technical drawings of complex machinery (Mellow 2018a).

The cost of hiring a professional illustrator will vary depending on the kind of work you are looking for and what you want to use the image for. An image made for a single poster can require a very different payment than an image that will be on a poster and in an open-access publication later. Expect creators to ask for higher fees for repeated reproduction or relinquishing their copyright (Mellow 2018a).

The overarching rule is to never ask for free work or to use original work without permission. The former is the height of rudeness (Mellow 2018a, 2018b) and the latter is most likely illegal. If your budget does not allow

FIGURE 5.1

Kelsi Rutledge and new species she described. (*Motor Trend*)

you to commission an original work, you might ask if there is older work that can be reused.

Outside creators can often bring in a whole new aesthetic to a project. Biologist Kelsi Rutledge worked with a professional photographer whose work also appeared in *Motor Trend* to take pictures of her and a new fish species she described (Rutledge 2019, Shiffman 2019b). Rather than staid photos of Kelsi in the lab with a preserved fish in a jar, the scientist and photographer went to the beach with a preserved specimen. The final photos (FIGURE 5.1) are more reminiscent of something you might see on a birth announcement. Rutledge said, "I think it's important to show that taxonomy can also be fun and that this research is important" (Márquez 2019).

An advantage of working with non-academic professionals, particularly graphic designers, is that because they are not specialists in your field, they will often force you to articulate ideas in new and unexpected ways. This exercise often ends up driving the research project forward (Khoury et al. 2019).

Conference registration

Conferences often make you submit a poster title and abstract (sometimes along with registration) months before the conference happens. This is where your work begins on a poster. Submitting so far in advance means that people often have projects that are not complete. You should not feel obligated to stick with anything you submit when you register for a conference. You can change your title or your poster content if you need to.

But this does not mean that you can do a slapdash job on your title or abstract, or that you should change these things on a whim. On the contrary, a few good decisions here can significantly increase the number of people who will want to visit your poster. The title, and sometimes the abstract, will appear in the program book. People will make decisions about whether they want to come to the poster based on that title and abstract. Sometimes, abstracts are published in proceedings or special issues of journals. These become citable pieces of the academic literature that can persist after the conference is over. You want these to be as high quality and as findable as possible.

Writing the title

Your title is the **most important part** of your poster. I cannot stress this enough. Nothing else even comes close. Your title is 80–90% of your communication effort (Ogilvy 1963, Olson 2014).

First, more people will read your title than anything else. They will read it in the program book. They will see it when browsing the hallways. Because they are expecting titles, they may even look at the titles before they look at images (Outing 2004).

Second, for people who do read your title, it is likely that your title is all that they will read. Experience with the web, advertising, and newspapers consistently shows that most people do not read content beyond a headline (Ogilvy 1963, Loranger 2015, Carson 2018). People share articles on social media based on headlines alone without reading the article (Loranger and Nielsen 2017). In fact, it's very likely that people are not even reading the entire title: they scan the first few words and give up if they are not encouraged by what they read (Outing 2004).

Third, your title can make a significant difference to whether those who at least read the title will bother to come talk to you and learn more.

Your title must make sense out of context (Ogilvy 1963, Loranger 2015, Carson 2018). Someone who cannot read the poster or abstract should still be able to understand what your project is about. Give them something they can take away even if they don't come to your poster. Make your title the take-home message. Make it a simple, short, declarative statement of your main result (Paiva *et al.* 2012, Lockwood 2016, Di Girolamo and Reynders 2017). Make a title that emphasizes a big picture, not minutia of your specific study (Jacques and Sebire 2010, Fox and Burns 2015). Try not to use technical terms, like species names of organisms in Latin (Murphy *et al.* 2019).

Try to keep your title as short as possible (Loranger 2015). Space on a poster is a limited resource and you want to make all of it count. A short title can be bigger and more visible on the poster and more easily shared on social media. Shortening titles is an editorial skill, but with practice you can recognize filler words and stock phrases (discussed in more detail in Chapter 17). For example, "The effect of X on Y" is weak. All it tells you is that you are looking for a relationship between two variables, which many research projects try to do. A title in that form can almost always be replaced by "X increases Y" (or decreases, as the case may be). It's shorter and more informative.

When I say the title should be as short as possible, the qualifier "as possible" is important. A title can be too short. The problem with very short titles is not their brevity but their ambiguity (Ogilvy 1963, Carson

2018). For example, the title "Cloning crayfish cell culture" used in the poster in FIGURE 1.2 is ambiguous because it is not clear if the crayfish or the cells are doing the cloning. (For the record, the crayfish clones itself.) Titles that include basic information like "who," "what," "where," "when," "how" get more attention, because those basic descriptors add clarity (Carson 2018).

For example, a title such as "Snake foraging" may be too short for a poster title. It does not contain enough information.

"Predation strategies of western diamondback rattlesnakes (*Crotalus atrox*) differ for mice (*Mus musculus*) and ground squirrels (*Spermophilus* spp.)" may be too long because it contains information that is irrelevant to the needs of the reader who is trying to decide whether to stop and talk to you. Scientific names for organisms are important for precision in the main text. You might have to use a Latin name for species that do not have a common name. But for familiar plants or animals, a common name may be all that's needed.

"Rattlesnakes ambush field mice but stalk squirrels" is closer to the right length for a poster title. It also conveys more knowledge about the findings of the project than the long title, by using precise verbs ("ambush," "stalk") rather than a more ambiguous phrase ("predation strategies").

If your research project uses a new tool or method, consider saying it in the title. New methods are always popular because the message "I can fix that problem you have" appeals to a viewer's self-interest (Ogilvy 1963). Again, the title of the poster in FIGURE 1.2 missed the opportunity to tell people that it was a new method.

If your research project contains anything that breaks any sort of record, that could be mentioned in the title. Superlatives like "highest," "furthest," "fastest," "deepest," or "oldest" attract attention (Carson 2018). Besides our natural fascination with extremes, record breakers are important from a research perspective because they set the bounds on what is possible. Record breakers often end up being the crucial data points that can strongly confirm or reject some hypothesis.

Do not make your title "clever" (Ogilvy 1963, Loranger 2015, Loranger and Nielsen 2017, Carson 2018). Avoid wordplay or oblique references. Don't use a multi-part title with colons (Jamali and Nikzad 2011). Conference goers are already overloaded with information and are scanning for relevant content. They do not have time to decipher or decode something obscure. If they do not understand what the title means, they will move on rather than stop to get clarification.

Similarly, don't make your title a question, trying to lure people in with a mystery.

The popularity of the suspense genre in books and movies has encouraged people to extrapolate this to maintaining interest when conveying new information as a salesman does when unveiling a new product. I think not knowing how something ends makes us apprehensive; it prohibits us from understanding how something was done while we frantically try to guess how it might end … While suspense has its place, it does tend to induce anxiety, which is probably not an optimum state for receiving new information. (Wurman 2001)

People are not at an academic conference because they want to solve a mystery. They can read a Sherlock Holmes or Phryne Fisher mystery somewhere else. If you want to emulate a fictional detective, be television's Lieutenant Columbo. In *Columbo*, the story begins by showing the murderer committing the deed. The fun comes not from "Whodunnit?" but from "How's he going to prove it?" Tell your viewer the result first, in your title, then use the body of the poster to prove it.

If there are any common key words that people might search for, include them in the title if you can. There will be a trade-off between keeping the title brief and making it more likely to pop up when someone searches for their favorite key words.

It is probably not a good idea to submit the first title you write. Single sentences are hard to write, and you will probably need to go through many iterations before you hit the "sweet spot" of informative but short. Advertiser David Ogilvy wrote at least sixteen headlines for every advertisement (Ogilvy 1963). Headlines made such a huge difference for sales that he could not afford to take the risk of using the first one he wrote.

If your title must be generic because the abstract deadline is so far from the conference that you do not have a clear idea of the main results, submit with the most specific title you can. Consider rewriting the title for the actual poster.

Writing the abstract

Most conferences will also ask for an abstract at the same time as the title. This can be challenging. When you are writing an abstract for a journal article, it will often be the last thing written, so the abstract can be very precise. In contrast, because abstracts are submitted well in advance of the conference, it's often the *first* thing to be written about the project. The project may be unfinished, and sometimes you may have to be a little vague.

But it is critical that you make your abstract as complete and as precise as possible. Conference abstracts are sometimes published, making them part of the scientific record. Because many projects presented on conference posters are never published (Scherer *et al.* 2018), an abstract may become the only record of some finding (Foster *et al.* 2019).

An abstract should follow the same four-part structure as an academic article: an introduction (why are you doing this and what is the question?), a very short description of methods (how did you try to answer it?), the results (what did you find?), and a discussion (what does it mean and how does it advance our knowledge?). Do not mention peripheral or tangential experiments. You want to end with a single point that nails your claim.

Because conference presentations are often works in progress, it's useful to try to signal in the abstract how far along in the project you are (Bastian 2018). If you are at the early stages, describe it as a proposal. If it's completed, mention the sample size.

If there were key words that didn't fit into your title, try to include words that can help people find your abstract (Bastian 2018). For example, in biology, if you used only a common name of a species in the title, use the Latin name of the species in your abstract. And think about related concepts. Someone who does research on pythons might be willing to come to a presentation on boa constrictors, so you might want to include the word "snake" in your abstract.

Include any identifiers associated with the project, like a registry number for a clinical trial or a grant number from a supporting agency (Bastian 2018, Foster *et al.* 2019). Speaking of grants, some conferences will ask abstracts to include a statement if there is some potential conflict of interest. For example, if your project is about obesity, and you were supported by a company that has ties to the food industry, you should mention that.

Proofreading

Because the title and abstract are the most likely parts of the project to be archived, check your writing carefully before submitting it to the conference organizer. Unlike journal articles, most conference abstracts are printed or archived "as is" and do not get the benefit of professional copyediting and proofreading.

I speak from experience when I say that these errors, while they may go unnoticed by many, will haunt you every time you think of that meeting. I have mangled species names in titles twice (Faulkes and Varghese 2004, Faulkes 2007). In one, the name I wrote for a spiny lobster, *Palinurus*, should have been *Panulirus*. (In my defense, both of these are correct names for spiny lobsters. *Panulirus* was created by forming an anagram

FIGURE 5.2

Poster abstract with
mistake in the title.

P3.73 FAULKES, Z.; The University of Texas-Pan American;
zfaulkes@utpa.edu

**Motor neurons involved in escape responses in white
shrimp, *Litopeneaus* setiferus**

Many decapod crustaceans escape from predators using tailflips,
which are generated by a set of giant neurons. In crayfish, medial
and lateral giant interneurons make monosynaptic connections
with abdominal fast flexor motor giant neurons (MoG) and
polysynaptic connections to other fast flexor motor neurons.
Shrimp are more basal decapods than crayfish, but shrimps'
escape circuit has features that are arguably more sophisticated
than crayfish: shrimp have myelinated giant axons with the faster
known conduction velocities in the animal kingdom, and the MoG
axons are fused. It is unclear what adaptive significance these
differences might have, as neither the escape responses nor the
neurons responsible for them are as well described in shrimps as in
crayfish. The abdominal fast flexor motor neurons of white shrimp
(*Litopeneaus* setiferus) were examined by backfilling with cobalt
chloride and nickel chloride, which showed that white shrimp have
among the smallest number of fast flexor motor neurons known in
decapods. Like most other decapods, the fast flexor cell bodies are
found in three clusters, but white shrimp have one or two fewer cell
bodies in each cluster than crayfish. The white shrimp MoG cell
bodies, rather than being round and filling uniformly like other fast
flexor cell bodies, have an irregular shape and a variegated
appearance, suggesting that the MoG cell bodies may result from
fusion of many smaller cell bodies. The previously reported fusion
of MoG axons in the nerve cord was confirmed. All fast flexor motor
neurons appear to be myelinated in the nerve cord, but not the
ganglion. The smaller number of neurons, myelination, and MoG
fusion in white shrimp suggest that shrimp tailflipping is more
coarsely regulated and less variable than that of crayfish, and that
duplication of motor neurons may have been important in the
evolution of decapod crustacean escape responses.

of *Palinurus*, which has frequently caused confusion, including my own.)
In the other (**FIGURE 5.2**), *Litopenaeus* became *Litopeneaus*. (That was
careless, and I have no excuse for it.)

Besides embarrassment, these mistakes make it more difficult for people
to find your work. Entering a misspelled name like *Litopeneaus* will
usually make a search engine say "Did you mean *Litopenaeus*?" and show
results for the more common correct name, but you cannot rely on this.
It takes effort to find misspelled things on the internet.

Before submitting your abstract, leave it alone for a day or more, then
proofread it. Rereading your text after a break can feel like reading it for
the first time. This is time to check for misspellings, missing words, and
extra words.

Many word processors will highlight words not in their dictionary.
Many people seem to distrust spell checkers or get so used to them that
they ignore warnings. Checking words with a red underline can avoid
"own goals" that can be easily fixed. But do not trust autocorrect too
much, because real words in the wrong context can look even worse than
made-up words. ("This person doesn't know the difference between 'its'
and 'it's'? Amateur.")

You have become so accustomed to looking at the way your text looks that
changing the appearance of the text can help you catch errors. Printing a
page instead of reading it on a computer screen is often enough. Changing

the typeface is a good way to find typos, because the new letterforms stop you from seeing what you expect rather than what is there. Changing column width is useful for finding repeated or missing words that you overlooked because they were at the ends of lines.

Reading text backwards, from the last word of the article to the first, is another trick proofreaders use to get them to focus on the text that is there, not what they expect to be there.

Reading text out loud is great for finding missing words, particularly small words that are common and we assume are there. Reading out loud also exposes convoluted sentences that people write but would never say.

While all these tricks are great to get you to look at your own text with fresh eyes, there is no substitute for a different pair of eyes. Different people write, edit, and proofread texts in professional publishing for exactly this reason. Ask other people to read your text. Reading an abstract will not take long, so it should be an easy favor to ask.

If you cannot give your abstract to anyone else, try to make several passes through it, looking for a different kind of problem each time. For example, you might do one pass for spelling mistakes, one pass for grammar and wording, and one pass to check numbers and names.

Planning the poster

You've submitted your abstract, and now it's weeks or months later, and it's time to start planning your poster. **This is the stage where most posters fail.** This is the point where you make a lot of important decisions, such as what you want to put on the poster and to leave out. Forcing yourself to decide now what data you will show and what you want the poster to look like will improve the content and appearance of your poster and make creating it faster and more efficient.

Read the instructions

Well-organized conferences will have clear information for their presenters. For poster presenters, there is at least one piece of information that you **must** find and pay close attention to. How big can your poster be?

Surprisingly, a printing company that specializes in conference posters reports that posters being the wrong size is the **most common problem** the staff encounter. People will ask for a poster that is a rectangle measuring 36 by 48 inches (915 × 1,220 mm) but send a file that is for a poster that is a square measuring 48 by 48 inches (1,220 × 1,220 mm). Apparently, people do not realize that the defaults of their software programs may not automatically match the proportions of the conference.

My own experience supports that mistakes on poster size are common. At one conference, I went through the conference halls and looked at every poster. Out of 400 posters, 4.5% of them did not fit on their poster boards (Faulkes 2010). There were probably more presenters who didn't read the instructions but made posters that happened to fit the space by coincidence.

None of the posters at that meeting was as obviously mis-sized as that shown in FIGURE 5.3, though. It does not matter whether this poster does a lot of the detail work right. It does not matter how good the layout is, or how good the typography is, or whether the color scheme is consistent and pleasing to the eye, or whether there is enough white space. None of that matters. The authors of this poster doomed it when they got the page size wrong. A poster that does not fit on its poster board is as rude and amateurish as giving a talk that goes badly over time.

FIGURE 5.3

Poster whose creators may have confused length and width. (Ewan McNay)

Competitions provide even more incentives to read guidelines. Many conferences have "best poster" competitions. If you are eligible, entering the competition is a good idea. Being a contestant guarantees an audience if nothing else, because the judges must look at your poster.

If you are entering a poster competition and want to win, find the criteria the judges use, if possible. Most competitions have multiple judges, but each of them has their own opinions. To make judging easier, there is usually a standard scoring sheet that judges use to assign points for different elements of the poster. Usually, the scoring system gives more points for the scientific content of the poster than for its visual design. But the judging criteria are usually more detailed than "75 points possible for the science, 25 points possible for how good the poster looks." The poster and the speaker's presentation of it are evaluated by multiple criteria. Sometimes there are as many as twenty. For example, there might be points for having an explicit hypothesis. They might be points for references. They might be points for an abstract. This means that those specific elements must be on the poster. Two examples are shown in FIGURES 5.4 and 5.5.

If you want to win the poster competition, use that judging sheet as your guide for making the poster. Make sure you hit every single criterion, regardless of whether this book says "That's a stupid thing to put on a poster and you shouldn't do it."

POSTER JUDGING SCORE SHEET

Presenter: _____
Date/Time: _____
Location: _____

Judge's Name: _____

Please CIRCLE your score in each category

Category	Outstanding	Very Good	Good	Fair	Missing
Abstract	5	4	3	2	0
Originality / Creativity	10	8	6	4	0
Hypothesis / Objectives stated	20	16	12	8	0
Experimental Design	10	8	6	4	0
Implementation of Design	10	8	6	4	0
Validity of Conclusions	10	8	6	4	0
Graphic Quality	10	8	6	4	0
Answers to Judge's Questions	10	8	6	4	0
Quality of Writing	10	8	6	4	0
Bibliography	5	4	3	2	0

Your Initials: _____   TOTAL POINTS: _____

PLEASE RETURN SHEETS TO PROGRAM OFFICER AS SOON AS POSSIBLE!

FIGURE 5.4

Sample score sheet for poster judging.

FIGURE 5.5

Sample score
sheet for poster
judging (different
conference than
Figure 5.4).

Judge's Score Sheet – Poster Contest

Student Name: _____   Poster Number: _____
University: _____

Criteria	Excellent	Good	Fair	Poor	Absent
Abstract	5	4	3	2	0
Appearance and flow					
Overall appearance/organization	5	3	1	0	-
Color is balanced	5	3	1	0	-
Print/font easy to read from a distance	5	3	1	0	-
Correct spelling and terminology	5	3	1	0	-
High quality graphs, figures, etc.	5	4	3	2	0
Introduction					
Introduction to topic/problem	5	4	3	2	0
Concise background information	5	4	3	2	0
Clearly stated objectives and/or hypothesis	5	4	3	2	0
Materials and methods					
Appropriate methodology to address problem	5	4	3	2	0
Concise explanation of procedures	5	4	3	2	0
Appropriate statistical procedures used	5	4	3	2	0
Results and discussion					
Clear and concise presentation of results	5	4	3	2	0
Clear reference to images, tables, graphs, etc.	5	4	3	2	0
Data collected compared to previous research	5	4	3	2	0
Conclusions					
Conclusions clearly support or refute hypothesis	5	4	3	2	0
Conclusions supported by results	5	4	3	2	0
Future research					
Clearly defined goals for future research	5	4	3	2	0
Presenter interaction					
Answers questions correctly and concisely	5	4	3	2	0
Enthusiasm/professionalism	5	4	3	2	0

Total Points: _____
Final Score: _____
Rank _____
Comments (Make additional comments on back if needed):

Digital posters

Screens are everywhere in our lives, and some conferences are moving to giving presenters a screen instead of a poster board to show posters. At some conferences, all posters are handled this way, while at others there is a limited number of screens slots, but most posters are still on boards. Some conferences are now happening entirely on social media, where there is only a digital format.

If you have the option for an electronic poster, there are pros and cons to consider.

On the plus side, the "killer feature" for digital posters is video and animation. Poster presenters have struggled for years to present information that is best understood as moving images. If your data would benefit from this, you should consider a digital poster. (Chapter 11 talks about how you can bring video to a poster session even if you have a paper poster.)

Because digital posters are uploaded, they are more convenient for traveling and less prone to disaster. In conversation with organizers, I learned that one conference attendee had a poster stolen! Why anyone would steal a conference poster is beside the point. The point is that the presenter had the poster saved somewhere as a portable document format

(PDF) file and was able to present the work as a digital poster. The crisis was averted and the poster thief thwarted (assuming the poster was stolen to stop the presentation). On a related note, a digital poster is more likely to be archived by the conference organizers, whereas you may never see a paper poster if you miss the session unless the presenter decides to archive it somewhere.

Digital posters are more likely to look exactly the way a presenter expects. Posters created on screens will look pretty much the same on a different screen. Converting a screen to ink on paper is a finicky business, and posters may look different than presenters expected or wanted. (More on this in Chapter 22.)

On the minus side, while paper is a relatively uniform standard, digital formats are not. Different providers of digital posters will have different formats they support, and different features that you can exploit. How large is the screen compared to a paper poster board (if available)? If the video screen is smaller, it may be more difficult to present than a larger paper poster. Does the digital poster format support video? What kind of video file formats does it support? What is the resolution of the screen? Can you zoom in to a section of the poster?

When does a digital poster have to be submitted to the conference organizers? If it must be uploaded well in advance of the conference, there may be time lost that could be spent working on the poster. (And academics are notoriously "last possible minute" people.)

For digital posters, the relevant measure may not be size in inches or millimeters, but file size. For example, some online conferences have been held on Twitter, which limits images by size. The service has changed its file size limits multiple times and could do so again.

Finally, there is a certain security in paper's low-tech approach. "Although paper can be lost, it almost never malfunctions" (Bakeman and Gottman 1986). Paper never needs a power cord or a reboot.

Budgeting time

A common attitude in science is that only collecting data and writing grant proposals or papers is real work. Spending time on design is viewed suspiciously, and even people interested in design will say things like "Grad students shouldn't be spending two hours fussing with arranging stuff on their poster" (Lucas 2019). But it is extremely unlikely that you can produce a decent poster in that amount of time. For one, you are probably generating text and figures that will go onto the poster long before you settle on a layout.

If you've done posters before, or only want a competent poster, not a better one, you can probably knock out a poster in an afternoon. Maybe even two hours. But if you want a better poster, you need to budget time for planning, laying out, and refining the poster.

Work backwards to figure out when you need to start working on your poster. What day are you going to leave for the conference? You want to have the printed poster in your hands the day *before* you leave. How are you printing the poster? If you are sending it off to a commercial printer, you probably need to give them the computer files at least a day before you need it, so that they can print and ship it. And it might take several days, particularly if there's a weekend or a holiday in there.

Then think about your workload. If you're like most academics I know, you have a lot on your plate, and you can't devote entire work days to designing your poster. You're probably going to lay it out in bursts of maybe half an hour here, a couple of hours there, over several days. There are advantages to this, because it means you can put a poster away for a day or two, sleep on it, and look at it with fresh eyes the next day.

As you'll see as you go through this book, making a good poster is about relentless tweaking and tinkering. You make fine adjustments, readjustments, throw the whole thing out and start again (hopefully not often), make more revisions, a few more tweaks, and so on.

I suggest starting your poster at least ten days before you plan to leave. But the more time you can give yourself to work through the design process, the better.

Chapter recap

- Have the hard conversations about credit with your colleagues early and often.

- Titles and abstracts may have to be submitted months before the meeting, which can make writing them challenging – but they are the most important thing you will write.

- Titles should be simple, declarative sentences that stand alone.

- Abstracts are often the only archived results from the poster.

- The biggest problems happen early. Read the instructions!

Narrative thinking

Design is driven not only by what you want to achieve (as set out by the design brief) but by your content. If you do not understand your content, all the design advice in the world will not help.

"How could I not understand my own project?" You probably do understand your project in a way that makes sense in your head. But you might not be able to articulate that to other people. The "curse of expertise" is that once you have gained knowledge, you forget what it's like to be ignorant of certain information, making it difficult to explain subjects to novices (Hinds 1999, Wieman 2007, Fisher and Keil 2016). It's easy to get so invested with the details of the project that you might not have worked out a "big picture" that you can quickly relay to others (Serrell 2015).

The training and collaborative structure of many research projects work against people understanding the context of their project. Posters are often made by students under the supervision of a more senior researcher (the "principal investigator," as they are called in the land of grant proposals). The senior researcher often has a long-term research plan that spans years, if not decades. Those long-term research plans make for a big pie, but new students joining the lab may only get one small slice of that pie. Those students may have very fuzzy impressions of where their projects fit into the larger whole.

Given the parameters outlined in the design brief (see Chapter 4), it stands to reason that good posters are concise. You need to figure out what to say that will get and keep someone else's attention. The most powerful and effective way to do that is to make sure your content has a strong narrative. But "narrative" is often misunderstood, because people think it is a synonym for "storytelling." When people hear "storytelling," they tend to think of fiction and children's bedtime stories. Researchers recoil at the slightest hint that they should ever present something that has any connotation of not being true. Researchers are loath to think that their hard-won conclusions could be as facile as a mere story. But narrative does not mean untrue. Narrative does not mean stupid. Narrative is more general than storytelling.

Defining narrative

A narrative is a series of events that occurs in solving a problem (Olson 2015).

Narratives engage people more than almost anything else. We tend to think of problems as bad, but problems are one of the things people love most! We spend much of our free time voluntarily creating arbitrary problems to solve (games and puzzles), or we watch other people solve problems (most movies, television series, and books). Viewed through this lens, it is instantly clear that most academic research has a narrative, because most academic research is performed because there is a problem that needs solving. It may not be a practical problem. It may be a highly theoretical and arcane problem. Whether a hypothesis is true – or not – is a problem, solved by testing it by gathering evidence.

But many researchers have a surprisingly weak sense of narrative, if they are not outright hostile to it. Repeatedly, when someone brings up narrative or its sibling, storytelling, someone will grumpily announce that this is bad for science. All you should need are a coherent hypothesis and an appropriate experimental design. What they fail to realize is that a hypothesis and an experiment constitute a narrative. The hypothesis states a problem and the experiment uncovers the underlying causal relationships. Cause and effect are at the heart of narrative. "The queen died and then the king died" is not a story because there is no causal connection between the two events. "The queen died and then the king died of grief" *is* a story, because there is a cause-and-effect relationship.

Academics have been trained to value content above everything else. Scientists will say "Show me the data" as if it were a mic drop. Even worse is the belief that "the data speak for themselves." But data never speak for themselves. Data always need advocates to place them in context and to convey importance. Data in isolation are just curiosities.

Researchers are also trained to be conservative. Not socially conservative, but conservative in the sense that they are restrained in their interpretation of their results. Academics are trained to look for exceptions and alternative explanations, because if they don't, their colleagues will, and being on the receiving end of those critiques is not pleasant. Researchers qualify, hedge, and insist that things are not simple.

For these and other reasons, many poster presenters do not understand what the narrative for their own research is, or they are unable to summarize it in a concise way.

Without narrative, scientists nuance themselves into irrelevance and oblivion.

Narrative helped Craig Bennett and colleagues create one of the most successful conference posters of all time, presented at the Human Brain Mapping conference in 2009 (FIGURE 6.1). Bennett and colleagues wanted to show common statistical problems in the field of brain scanning (functional magnetic resonance imaging, or fMRI) that could lead to "false alarms" in experiments. Bennett had scanned a dead salmon for other reasons but noticed what appeared to be brain activity in his deceased fish. It was a spurious false-alarm result, but that was the point. Bennett recognized the narrative potential of the fishy false alarm. He wrote, "If they would have been anywhere else the salmon would have been just a curious anecdote, but now we had a story" (Bennett 2009).

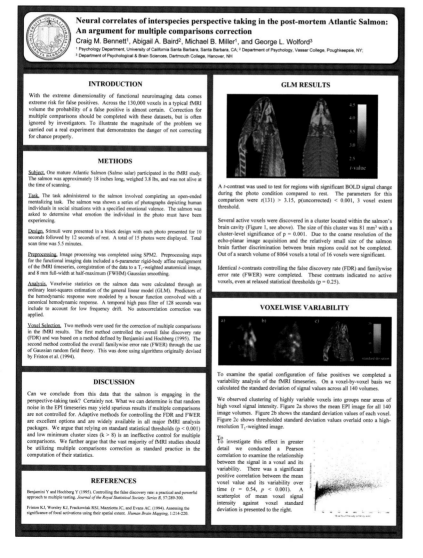

FIGURE 6.1

"Dead salmon" poster. The poster text notes that the salmon "was not alive at the time of scanning." (Craig Bennett)

The conference organizers almost rejected the poster abstract because they thought it was a joke (Bennett 2009). It was a joke, but it was a joke with a serious point. The poster design ignores much of the advice in this book, but the combination of a strong narrative, humor, and an important point for the research field made for a winner. The poster has been cited over 300 times (Bennett 2009, Bennett *et al.* 2009). Hundreds of citations would be excellent for any journal article, but that many citations for a conference poster is nothing short of astonishing.

Hypotheses

A standard question for many researchers when viewing a poster is to ask the presenter, "What is your hypothesis?" Not all research is hypothesis-driven, but it is so common in so many projects that it is valuable to understand how to state them. Many people do a poor job of articulating their hypotheses.

If asked, "What's your hypothesis?" don't start your reply with something like, "My question is …" A question alone is not a hypothesis.

Strong hypotheses have two components. First, hypotheses suppose some sort of a mechanism. Hypotheses are statements of cause and effect: "I think *this* causes *that*." Strong hypotheses incorporate a mechanism, whether implicitly or explicitly. Second, strong hypotheses make predictions. For example, "Striped tailfeathers of birds are sexual signals that evolved due to sexual selection and female mate choice" (cause and effect) suggests, "Males with more stripes on their tails will mate more often" (prediction).

"My control group will be different than my experimental group" is not explicit enough. Okay, you've learned the concept of the null and alternative hypotheses, which is useful for statistical analyses. But if that is your best prediction, the project is better described as "hoping" than as being informed by hypothesis.

The null and alternative hypotheses are a familiar example of competing hypotheses. But competing hypotheses that make different predictions is an even more compelling narrative (Platt 1964). "If hypothesis 1 is true, the test group will have a lower average than the control, but if hypothesis 2 is true, the experimental group will have a higher average than the control" makes it less likely that your results will be ambiguous and hard to interpret. If your project uses this strong inference framework, mention both hypotheses and how their predictions differ.

"And, but, therefore" sentences

A powerful tool for creating a concise narrative is a single sentence containing three words: "and," "but," and "therefore" (Olson 2015, 2019a). This is the ABT format.

If you can describe your work in an ABT sentence, you have an unfair advantage in the competition for attention. I've listened to many peoples' description of their research, given it back to them paraphrased in an ABT sentence, and seen their faces when they realize, "Oh, that really does sum up the work." Minutes of convoluted description can be replaced with one sentence.

While narrative is not identical to storytelling, the ABT sentence resembles storytelling in that it creates a structure that might be familiar. You can think of it as a miniature, self-contained, three-act play.

The first two parts of the sentence are two facts, joined by "and." This is the exposition that sets up the situation or cast of characters.

The "but" explicitly identifies the problem to be solved, which is the heart of what defines a narrative. Here is where some of the similarities to more familiar storytelling might be obvious: "but" creates a conflict and, as the saying goes, conflict is the essence of drama.

The word "therefore" either resolves that problem, or at least indicates a way that the problem might be solved.

Before turning our attention to technical writing, let's use some examples. You can find this structure in pop songs:

> Hey, I just met you AND this is crazy, BUT here's my number, so [THEREFORE] call me maybe.

Here's Goldilocks and the three bears:

> A family of bears go out of their house AND leave the door unlocked, BUT a young girl wanders into the house, not knowing who it belongs to, and THEREFORE gets terrified when the owners return.

How about Robin Hood?

> Prince John is in power while his brother Richard is fighting in the crusades AND is taxing and oppressing the poor, BUT Robin and his band of outlaws rob from the rich and give to the poor, THEREFORE making the lives of the poor better.

Here's *Star Wars*:

> An empire has enslaved the galaxy AND is building an ultimate weapon, BUT the weapon's plans are stolen by rebels, THEREFORE bringing new hope to the galaxy.

The "dead salmon" poster mentioned above:

> We scanned the brain of a dead salmon AND saw what looked like brain activity BUT this is obviously impossible and must be a false alarm, THEREFORE we should correct for multiple comparisons to reduce the chance of spurious conclusions.

Obviously, there are more to these stories than just a single sentence. That's why storytelling is different from narrative. Storytelling is about the specifics of character and situation and setting and plot, none of which are necessary for narrative. But you can recognize how the "introduction, complication, solution" format creates a narrative. These examples should also make it clear that while I've been emphasizing the ABT structure in single sentences, the ABT structure is in play at higher levels, too. I already compared the sentence structure to a three-act play.

The word in an ABT sentence that turns facts into narrative is "but." It should be easy to determine a source of conflict if you are doing hypothesis-driven research: "We have this hypothesis, *but* it has never been tested, therefore we did this experiment." It can be as simple as a gap in knowledge: "We know this and this, *but* not this, therefore we tried to fill the gap."

There are other single-sentence formats to try to distil complex subjects down to an essential message. One is "You know …? / Well …" (Eng 2012). For instance, "You know how when you're 10 minutes away from the presentation of your life and your device has stopped working and you can't get tech support on the phone? Well, I support and promote open-source practices which will allow you to fix those problems yourself."

But it doesn't take much to realize that "You know …? / Well …" is a variant of the ABT. "You know …?" is a statement of a problem ("and" and "but") and "Well" is the consequence ("therefore"). The example above could easily be reworded into an ABT sentence: "You're ten minutes from a big presentation and your device isn't working, but you can't get tech support on the phone, therefore you need practices that let you fix those problems yourself."

Once you look, you can start seeing ABT structure in academic writing. Sometimes it is apparent in the start of an abstract (Billings and Schnepel 2018):

> Lead pollution is consistently linked to cognitive and behavioral impairments, yet [another word for BUT] little is known about the benefits of public health interventions for children exposed to lead. [THEREFORE] This paper estimates the long-term impacts of early-life interventions (e.g. lead remediation, nutritional assessment, medical evaluation, developmental surveillance, and public assistance referrals) recommended for lead-poisoned children.

While academic writing typically does not teach the ABT structure in name, it exists in most scientific papers in practice. Technical papers start by laying out existing knowledge ("and"), identify a gap in knowledge ("but"), and run experiments to fill the gap ("therefore").

There is a real appetite for narrative in academic publishing – and it does not always have positive consequences. (Narrative is like the Force: it has a light side, it has a dark side, and it holds the universe together.) The most renowned scientific journals in the world, *Nature* and *Science*, are often criticized for wanting to publish research that has unexpected findings. More generally, there's been a long-term increase in the number of scientific papers that use words like "novel" (Vinkers *et al.* 2015). Why "novel" and not "new"? Because "novel" implies something unexpected in a way that "new" does not. Conversely, research that sets out to replicate previously published findings is more difficult to publish, particularly if that research confirms the previous result.

Surprising findings feed our desire for narrative. "We thought we knew this AND previous research supported it BUT new research contradicts that and THEREFORE everything we knew was wrong." There is a saying that movies are as good as their villains; narratives are as compelling as their problems. Confirmation and replication are critical for scientific confidence and integrity but make for poor narrative. "Demonstrate something a second, third, and fourth time so that we're sure" is not a compelling problem.

Now that you are familiar with the ABT structure, it's helpful to look at a couple of common patterns of explanation that fail. Both are common among academics (Olson 2015, Olson 2019a).

As mentioned before, many academics start trying to explain their projects by identifying a research area, rather than a problem. They list a set of facts. They give the impression that their project is adding one more fact to the list. This list-of-facts approach is "and, and, and" (AAA for short). It's not a narrative, and it is boring. Dictionaries list many facts, but people do not read dictionaries because they are engaged by them. They read dictionaries out of necessity, not pleasure, and return them to the proverbial bookshelf as soon as they solve the one problem they had.

At the other end of the spectrum are people who add so many caveats and qualifiers and nuances that they end up emphasizing all the exceptions, so you are never sure if what they are talking about applies to this situation. Because these explanations identify problems, they can technically be called narratives, but they are not strong narratives because they are confusing. This complexity can be summed up as "despite, however, yet" (DHY for short).

At this point, I have no doubt that many readers will want to say, "Lists of facts can be important!" and "Complexity and nuance is important!" Yes, they are important in the grand scheme of developing reliable knowledge. But remember the design brief and the goal of posters (Chapter 4): they are to promote conversation between you and other people. Narrative (and the ABT structure) promotes engagement in a way that listing facts (the AAA structure) or complexity (the DHY structure) do not.

No topic or project is stuck in any of these formats. You may have seen the same topic covered in different ways. You may have thought the American Civil War was boring when you learned it in school (AAA) but got sucked into a Ken Burns documentary about it (ABT). You may have though the global financial crisis was confusing (DHY) when you heard it on the news but understood it when you watched a movie about it like *The Big Short* (ABT).

The ABT sentence format is simple, but, as we saw with titles and headlines in Chapter 5, it can take many tries to get a single sentence that is just right.

Using narrative to plan

First, try to articulate a summary of your project. While an ABT sentence is a strong structure for this exercise, your summary might be based on a hypothesis statement, or even something more freeform. But because a poster has such limited space, building from a concise summary at the beginning is probably more fruitful than throwing in a lot of material and then trying to cut back.

At this stage, it is worthwhile trying to articulate that summary out loud, verbally. "You start writing, you start complicating" (Olson 2019b).

> You begin conceptualizing a presentation by figuring out your ABT. It's the first thing you assemble. It's your "narrative road map." It tells you what to keep in, and it tells you what to cut out. It tells you what "advances the narrative" (keep it in) and what is "off the narrative" or essentially a sidebar (cut it out). Once you've locked in your ABT, you then use it to "build out" your presentation. (Olson 2019a)

It's also worth asking any colleagues or coauthors to write their own ABT sentences for the poster. Here, the conciseness in an ABT sentence can make sure everyone agrees on what the narrative *is*:

> You need everyone to be working towards common goals – not wasting time on bad directions, but instead focusing all effort in the right direction. Knowing "the narrative" of your mission is central to achieving this. If you know the narrative, then you know what is

essential to advance the mission. You also know what is a waste of time and resources. (Olson 2019a)

If you and your collaborators are writing very different sentences, this is a sign of trouble.

Once you have worked out your narrative, big idea, or central hypothesis – call it what you will – refer to it constantly. Always be asking whether the material on your poster advances the narrative. Ask whether the data are relevant to your hypothesis. Ask whether the text is related to your big idea. If it is not, edit it out and save it for another project.

Chapter recap

- Narratives are events that occur during solving a problem.

- Cause-and-effect relationships are key to both narratives and hypotheses.

- Many narratives can be encapsulated in an "and, but, therefore" (ABT) sentence.

CHAPTER 7

Visual thinking and graphic design

Ideas without images are forgotten. (Wurman 2001)

Posters are a visual medium. Academics are used to thinking abstractly using words, and it can be challenging to present your work visually. But many complex problems can be presented using images, and this often makes them easier to solve. "The real goal of visual thinking is to make the complex understandable by making it visible – not by making it simple" (Roam 2013).

Dan Roam argues that there are six basic ways to show something, and you can recognize which you need by the kind of question you hear (Roam 2013):

- If you hear a name – a "who or what" – you need a portrait. This is not necessarily a realistic or detailed portrait like a painting or a posed photo. A stick and ball chemical structure is a "portrait" of a molecule. A smiling emoji can be a portrait.

- If you hear a number – a "how many" – you need a chart or graph. A bar graph is a simple example.

- If you hear a location or a list – a "where" – you need a map. Again, this need not be a literal cartographic map. Anytime you talk about something "above," "below," "closer," or "overlapping," you have the potential to create a map. Examples include concept maps, pedigrees and phylogenies, org charts and Venn diagrams.

- If you hear a history – a "when" – you need a timeline. "Time" is one of the most common variables shown graphically (Tufte 2001).

- If you hear a sequence or process – a "how" – you need a flowchart.

- If you hear some complex combinations – a "why" – you need a multi-variable plot, like a scatterplot.

"If the hypothesis is correct, then the measurements of the treated group will be significantly higher than the control group."

PREDICTION

CONTROL GROUP

TEST GROUP

FIGURE 7.1

Verbal versus visual descriptions of predictions.

Some of the best practices for making these graphs will be discussed in Chapter 9.

There are many variations of these, and there are probably data visualizations that may not neatly fit these categorizations. But this taxonomy goes a long way to answering, "How can I show things in a visual way?" In posters, the results are often depicted visually, but there are many other opportunities to turn sentences into pictures. For example, in an introduction, you might have an image showing the subject of your research instead of just naming it. Instead of stating a prediction arising from your hypothesis in a sentence, it could be shown as a pared-down graph. In FIGURE 7.1, the graph is drawn in sketch form to signal something preliminary instead of completed.

Verbal descriptions of methods might be replaced by flowcharts. Surprisingly, methods are rarely turned into images in the same way results are, despite the clear advantages of working graphically (Bastian 2019).

But knowing "I can use an image" is not enough. You need the concepts of graphic design to maximize the effectiveness of what you show.

Pictures and graphics

Posters are a visual medium, but not all visuals are created equally. Scott McCloud (1993, 2018) created "the big triangle" to map out different kinds of graphics and images (FIGURE 7.2). Start down in the bottom left corner, with "reality."

That bottom left corner represents images that are faithful representations of real objects. These images are instantly recognizable and understandable. People know real objects. You don't need practice to recognize real objects, or representations of real objects. You don't have to receive specific training on how to recognize a photo of a face.

But images don't have to look like objects in the real world. Images can be abstract. There are two ways to move away from the "reality" corner

FIGURE 7.2

The "big triangle"
of graphic
representations.

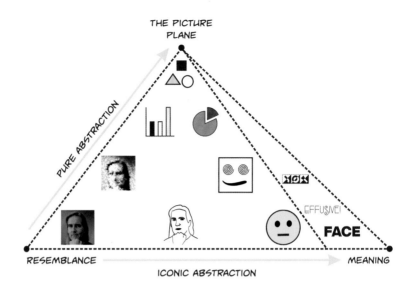

through abstraction. The first is by pure abstraction towards the "picture plane," where *both* resemblance and meaning are lost. The second is iconic abstraction, where the *meaning is kept* even as resemblance is lost. Words look nothing like the objects they represent, but they mean things. The presentation of research usually hugs the bottom edge of the big triangle, because most research is intended to say something meaningful about reality. But some graphs and charts are more abstract. They drift up towards the picture plane.

As you move away from that bottom left corner of the big triangle, whether by going up or by going to the right, you make it harder for a reader to understand what is going on. That line dividing the triangle in two is where the difficulty ramps up enormously. It marks literacy. People go through years of education to push past that dividing line so they can explore the bottom right corner of the triangle. But academics spend so much time there, they become very comfortable and forget that *reading is hard*. Forcing people to read if they don't have to is an unnecessary barrier.

The big triangle shows why photographs are so valuable on posters. Photographs are immediate. Photographs show real objects in a way that needs no specialized training to understand. Words need reading fluency. Graphs need interpretation. Tables need both literacy and interpretation. Advertisers learned decades ago that ads with photographs sell more product than those with artwork (Ogilvy 1963). Pictures of people attract more attention than pictures of almost anything else (Djamasbi *et al.* 2010).

A big, high-quality photo that relates to your topic anchors your poster to reality and gives people an entry point. In biology, one of the simplest entry points can be to have a big picture of the organisms you are working on. If you can possibly find a relevant photo, consider using it somewhere. In mathematics, a photograph of a mathematician who wrote a theorem could accompany the theorem itself.

That said, there are many cases where diagrams can show things more clearly than photographs (Seddon and Waterhouse 2009) or any other kind of literal representation. For example, maps of subway systems rarely show realistic directions and distances between stations. In descriptions of anatomy, technical illustrations are still preferred over photographs, for example. Besides clarity, diagrams can be more evocative. Because they tend to be stylized, they can be more flexible.

Perhaps the biggest downside of diagrams is that many poster creators do not feel comfortable in their ability to draw. And illustrations will typically take longer to create than a photograph.

Similarity and contrast

Graphic design often seems to boil down to two contradictory directives:

1. Make things look the same.
2. Make things look different.

I know that put that bluntly, it sounds crazy. But a little reflection shows that both are necessary. If everything looks the same, you have something dull. But if no two things are the same, you have something chaotic. Good design finds balance between those two extremes. Think of similarity and contrast as the yin and yang of graphic design (FIGURE 7.3).

Designer Robin Williams (2004) says that four main rules in graphic design are **repetition** (makes things look the same), **alignment** (make things the same), **contrast** (make things look different), and **proximity** (bringing things together makes them more the same and moving things apart makes them more different).

THE SYMBOL OF YIN AND YANG USES *SIMILARITY* IN SHAPES BUT *CONTRAST* IN COLOR AND DIRECTION!

FIGURE 7.3

Yin and yang symbol.

At some level, we know this intuitively. Books have black ink on white pages (contrast). But each line of text is aligned to the margin (similarity) and has equal spacing between each line (similarity). If you want to emphasize something, like a section heading or a word, you change its size, weight, or type (contrast). But the same method of indicating emphasis is used every time: chapter headings are the same weight, type, and style as each other (similarity), while they are much different from the main text (contrast).

If similarity and contrast represent two ends of a spectrum, different styles emerge as you go from one end of the continuum to the other. If you have very little variation in text, color, and positioning, you can create something that you might call subtle – or boring, depending on how well you pull it off. If there is a huge among of variation in text, color, and positioning, the look might get called exciting – or gaudy, depending on how well you pull it off (FIGURE 7.4). If you have few differences in text, color, and position, but those differences are deliberate and big, you might have a look that people call bold. The level of familiarity a viewer might have affects the assessment of style, but there's no doubt that the degree of resemblance between elements on the page plays as a big role.

Contrast plays a huge part in human perception, and therefore is integral to any sort of visual communication (Bang 2000). Here are some ways that elements on a page can contrast (Landa *et al.* 2007, Visocky O'Grady and Visocky O'Grady 2008):

Color – Color is perceived quickly even from long distances. Many people whose eyes cannot focus correctly can still make out colors of objects easily. Contrasts could include black versus white, bright versus dull, warm colors versus cool colors.

FIGURE 7.4

Examples of subtle, gaudy, and bold typography.

Subtle

Neutral colors like browns, grays, pastels
Little variation in size, color, or type
Common typefaces in a book weight

Gaudy

❖ **INTENSE PRIMARY COLORS**
❖ *No consistency in size, color, or type!*
❖ Everything small! Tiny fiddly details!

Bold

High-contrast colors
Differences are few but large
Large type (not necessarily bold weight)

Size – While we might normally describe objects on a page as big or small, line sizes might be described as thick, thin, or hairline.

Shape – Objects might be made mainly of straight lines or curved ones. At a higher level, these might lead to shapes being called "geometric" or "organic." Typefaces are mainly grouped by shape using terms like serif and sans serif.

Position – Differences in position can manifest in many ways. One object might be aligned with other elements on the page, or not aligned at all. One object might be closer to others (proximity). Position can also refer to whether an object is horizontal, vertical, or diagonal, and to whether blocks of text are justified or ragged, centered or off-centered.

Texture – On a printed page, this usually refers to an appearance of texture rather than an actual difference when you touch the page. Texture is an illusion achieved by the use of shape and color. But it can be easier to describe an image as looking smooth, furry, or rough, rather than trying to break down all the individual components that create the appearance of texture.

All of these can be played with to direct readers and organize information (FIGURE 7.5).

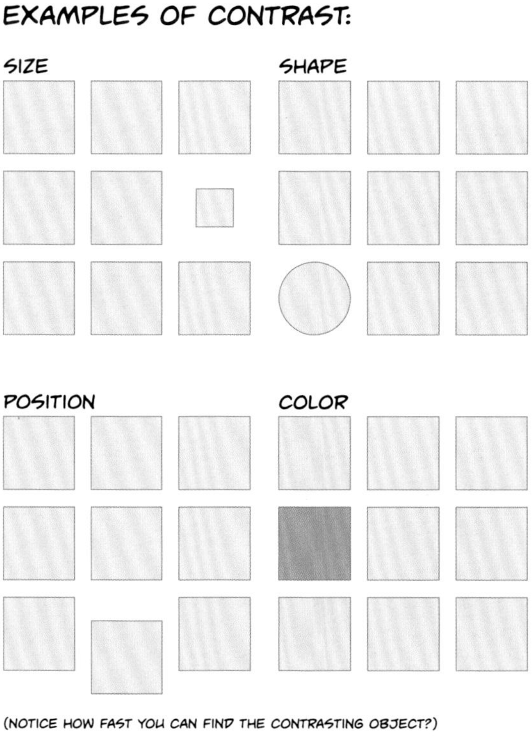

EXAMPLES OF CONTRAST:

FIGURE 7.5

Examples of visual contrast.

(NOTICE HOW FAST YOU CAN FIND THE CONTRASTING OBJECT?)

Visual hierarchy

Contrast and similarity are used to signal importance and direct attention. It is rare that everything on a page is equally important, and people expect cues to tell them the most critical information in your poster (Williams 2004). We pay attention to contrast, and treat contrasting items as important. We pay less attention to similarity, and treat similar items as related and less individually important.

Together, cues like these help to organize content by creating a visual hierarchy.

The first cue is contrast in **size**. Titles are bigger than headings, which are bigger than the main text, and the main text is often bigger than acknowledgments or references. Once you have decided how big your headings are, make them the same size everywhere on the poster. Similarly, an important results graph should be bigger than a minor picture to help clarify the methods. People look at larger items sooner, and for a longer time (Outing 2004).

In FIGURE 7.6, there are four maps showing species distribution models, each of which represents the results of one model. The authors considered one map to be more likely than the other three to predict the outcome, so it was given the most space on the poster.

FIGURE 7.6

Using size to emphasize importance.

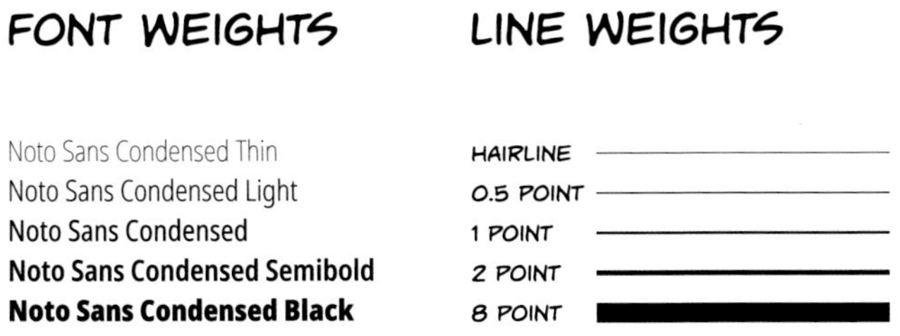

FONT WEIGHTS

Noto Sans Condensed Thin
Noto Sans Condensed Light
Noto Sans Condensed
Noto Sans Condensed Semibold
Noto Sans Condensed Black

LINE WEIGHTS

HAIRLINE	────────────
O.5 POINT	────────────
1 POINT	────────────
2 POINT	────────────
8 POINT	▬▬▬▬▬▬▬

FIGURE 7.7

Examples of differing font and line weights.

A size cue that is mostly relevant to type is contrast in **weight**. This is a little different than size. While people talking about the size of type normally measure its height, weight refers to the thickness of the lines. Bolding is the one form of weight that most are familiar with. Many typefaces come in a variety of weights (FIGURE 7.7), sometimes with names like Light, Black, Ultrablack, and so on.

Weight is also sometimes used to describe the thickness of lines or rules. A thick line has more weight than a thin one and a solid line has more weight than a dotted or dashed one.

A second cue is contrast in **position**. In a representational image, like a photograph or piece of art, the center of a page is usually the center of attention (Bang 2000). Magazines and newspapers usually place important material in the top center (Landa *et al.* 2007).

Grouping is another example of using a position as a cue. Things placed together tend to be treated together and considered "the same." Objects that are further apart are considered different. Many designs use this extremely effectively, by placing one item extremely far from all others, isolating it and drawing attention to it. In a graph, these points are called "outliers" and receive extra scrutiny because they seem to signal some sort of different cause generating them.

Positional cues become more complicated when text is involved. Experienced readers of English and many other languages know to read from the top down, and from left to right. When there is text on a page, elements at the top are considered more important than those at the bottom. Items further to the left are more important than those to the right (for example, indented text). Text that is centered within a grid is considered "higher" in the hierarchy than text that is left-aligned or justified. Every indent to the right lowers text's position in the hierarchy.

A third cue is **color**. Bright, primary colors tend to attract attention compared to neutral colors or grays. In addition to the hue, the contrast between one color and another can move an object up the visual hierarchy. A neutral gray that is mostly surrounded by red might seem more important because of the contrast.

FIGURE 7.8

Examples of
clear and unclear
visual hierarchy.

GOOD VISUAL HIERARCHY:

Title

Heading

Subheading

Body text body text body text Lorem ipsum dolor sit amet,
consectetur adipiscing elit. Aliquam porta elit et justo malesuada,
nec dapibus sapien egestas. Donec laoreet justo nec elit tincidunt
efficitur. Pellentesque vel egestas lorem. Morbi interdum, erat eget
aliquet eleifend, mauris turpis tempus risus, lacinia dictum tortor

CONFUSING VISUAL HIERARCHY:

Title

Heading

Subheading

Body text body text body text Lorem ipsum dolor sit amet,
consectetur adipiscing elit. Aliquam porta elit et justo malesuada,
nec dapibus sapien egestas. Donec laoreet justo nec elit tincidunt
efficitur. Pellentesque vel egestas lorem. Morbi interdum, erat eget
aliquet eleifend, mauris turpis tempus risus, lacinia dictum tortor

All these cues work simultaneously and in complex ways. For example,
a large object is generally higher in the visual hierarchy. But if that large
object is a pale color, it might have less visual weight than a smaller
object with intense color. A clear visual hierarchy uses these multiple
cues consistently. In FIGURE 7.8, both the position and size of the title
signal that it is the most important element in the top text sample. In the
second sample, the position of the title indicates it should be the most
important thing, but its small size gives a contradictory cue, suggesting
that the title is not important.

Don't use all the cues for emphasis at the same time. For example, you can
use size alone to call out your title. **Bolding** alone is fine for emphasis
in the main part of the text. But using all these tricks together at once
for emphasis is

TOO MUCH

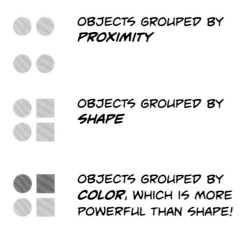

OBJECTS GROUPED BY **PROXIMITY**

OBJECTS GROUPED BY **SHAPE**

OBJECTS GROUPED BY **COLOR**, WHICH IS MORE POWERFUL THAN SHAPE!

FIGURE 7.9

Examples of cues that group content.

Besides defining importance, a clear visual hierarchy establishes **relationships** between related objects or concepts. By making two things the same font, size, or color, you indicate that those things are potentially related. Objects can be grouped by their position, their shape, or their color (FIGURE 7.9). And again, some cues are more powerful than others. Color is stronger than shape, for example (Bang 2000).

Another way of thinking about visual hierarchy is that it's a way to command and direct attention. I am reasonably sure that, when you flipped open this page, the last word of this sentence is the one you noticed first.

In FIGURE 7.10, the size, weight, and contrast of the sentence "You will read this first" overrides the cue of position, which would normally dictate that "You will read this last" be read first.

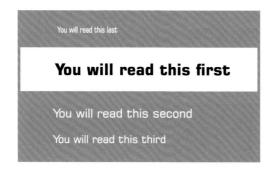

FIGURE 7.10

Example of visual hierarchy used to guide reading order.

White space

Novice poster designers who think that a poster needs to show all their information can get frustrated by the poster format because posters have finite physical space. Some researchers are not used to this constraint because they give more slide presentations than posters, and you can always add "just one more" slide.

People often try to solve the problem of limited poster space by trying to use all the space on the page. They make the margins small, the line spacing small, the pictures small, and the text small, all in the name of getting more bits of information on the physical page. Their rationale is that anything that doesn't contain information in the form of text, data, or images must be a waste of space. They undervalue empty space, also called white space (which, despite the name, can be any color).

But it is wrong to think that white space conveys no information. On the contrary, white space can convey several kinds of information.

White space is the single most powerful way of organizing a poster. As we have seen, position is a critical cue showing the relative importance of elements in the visual hierarchy, and the relationships between them. Because we judge whether pieces of content are related to each other by their proximity to each other, increasing the white space between two objects signals that they are unrelated. In comics, adjusting the amount of white space between panels can change the reading order of those panels (Cohn and Campbell 2015).

Large areas of white space can indicate an item's significance by removing visual competition. White space can direct people to entry points. Because we look for contrast, one of the most effective ways of drawing attention to "something" is to surround it with a whole lot of "nothing." A tree attracts no attention in a forest but becomes remarkable in a desert. Advertisers know this principle well. Large swaths of empty space are used in ads to signal high-end luxury goods, because the white space emphasizes the object and brands it as important.

In text, white space can signal the start of a new topic, by using indents to signal the start of a new paragraphs or a blank line to separate two paragraphs with no indents.

Besides providing informational cues, white space is an expected component in paper documents. Because white space doesn't draw attention to itself when used well, people often underestimate how much of a typical page it can take up. On a typical piece of letter-sized (or A4) paper, margins alone might take up about 37% of the page. And the percentage will increase depending on the spacing between the lines.

Posters that are chock-a-block with closely packed text are intimidating. Posters with lots of white space are welcoming. White space provides "breathing room" in a design, and many posters can be improved by moving elements apart to create more white space. This might require shrinking some things. You may need to use a smaller point size or a smaller picture, but the improvement will be worth it.

Entry points

The typical conference goer is like the little girl Alice who takes a trip to Wonderland (Carroll 1865). Alice is minding her own business, considering picking flowers, when she sees a White Rabbit. And when Alice watches the Rabbit take out a pocket watch, she is hooked and becomes determined to follow the White Rabbit wherever it goes. Because a White Rabbit with a pocket watch is remarkable and her curiosity about him is just that irresistible.

Every poster needs a white rabbit: an entry point for the readers, something that will draw them in when they had no intention of stopping at your poster. Poster presentations are often crowded, busy spaces with many people who have never met you and know nothing about your research. Your entry point, therefore, should be something that should be obvious. In other words, you need something that has a lot of visual weight. As discussed in Chapter 7, there are lots of ways to create that focal point.

What do you have on your poster that beckons someone to come closer? What on your poster might reach out to someone strolling by to lean in and have a second look?

Because people follow the Principle of Least Effort, entry points are necessary for people to connect with content (Trost 2009). People will look at headlines before reading a long story – if they read it at all. "Designers like to joke that no one reads copy" (Visocky O'Grady and Visocky O'Grady 2008). Pete Faraday (2000) noted that readers of web pages search for an entry point first, and suggested a number of elements that can act as cues for entry points. In order of importance, these are motion, size, imagery, color, text style, and position.

There are several kinds of entry points you can have on a poster.

Photographs almost always make the best entry points, for reasons the big triangle made clear (see FIGURE 7.2). People typically look at photographs first and more often than other elements in a page, particularly photographs of faces (Holmberg 2004, Djamasbi *et al.* 2010). But some photographs are still going to be unfamiliar to many viewers. A photograph of a slime mold in a petri dish might still be baffling unless you are at a slime mold conference. In cases like that, artwork or text might make a better entry point than a photograph.

Text might work as an entry point. For experienced readers, short blocks of text can be perceived almost as quickly as pictures. Headlines, section headings, and pull quotes (discussed later) can pop out and help people understand the content. Some research shows that people look at a headline even before they look at images (Outing 2004). It is unlikely that a graph can act as an entry point, because most graphs are too generic and practically anonymous. To have a chance of working as an entry point, the graph would need to be simple and well annotated.

A table will almost certainly *not* work as an entry point.

Pastiche

Do not seek to follow in the footsteps of the wise; seek what they sought.
(Matsuo Basho)

Having said that design is about making decisions, my next piece of advice might sound contradictory. An efficient way of designing your poster is to find bits of graphic design that you like and use them. This could be almost anything. A movie, an eggcup, an album cover, a skateboard, another conference poster, whatever.

This is not about mere imitation. This is about analyzing the original design, understanding what it was that appealed to you, and using that for your own purposes. If a design works in one context, it will probably work in other contexts. It can provide a certain level of "safety" that you don't get when you are trying to create a design that is entirely new and never before seen.

Game designer John Wick calls this "stealing the sentiment." Indeed, you see imitation all the time in games. Game mechanics cannot be copyrighted. They can be patented, but the trouble is such that most game designers don't bother. An innovation in gameplay in one game is often picked up and used in later games. Health bars in video games are now ubiquitous, but they trace back to a video game called Dungeon Buster (Macgregor 2018).

Academics might recoil at this suggestion, because it sounds like plagiarism. But imitation is the norm in many industries. In the fashion industry, clothing designs are not covered by intellectual property protection (Blakley 2010). The same with cooking: recipes are not protected. Similarly, the general "look" of something is not copyrighted in the same way a piece of text or a specific piece of art might be.

Posters can be inspired by the look of many things. Gavin Abercrombie (FIGURE 7.11) and Neil Cohn (FIGURE 7.12) stole the sentiment of comics. Even if you don't imitate the style of comics, posters and comics share so much in common that comics can provide some of the best lessons for poster creation.

For the poster shown in FIGURE 7.13, Jessica Stanton stole the sentiment of nineteenth-century newspapers. She reasoned that this was the time period when her subject, passenger pigeons, were on their way to extinction, making this an appropriate style to evoke on her poster.

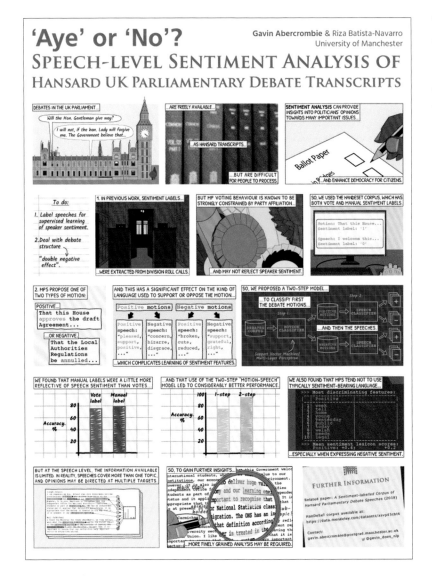

FIGURE 7.11

Poster design inspired by comics. (Gavin Abercrombie)

FIGURE 7.12

Poster design
inspired by comics.
(Neil Cohn)

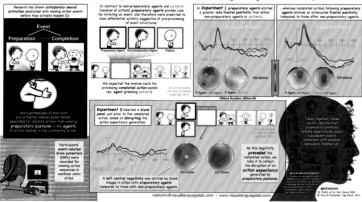

FIGURE 7.13

Poster design
inspired by vintage
newspapers.
(Jessica Stanton)

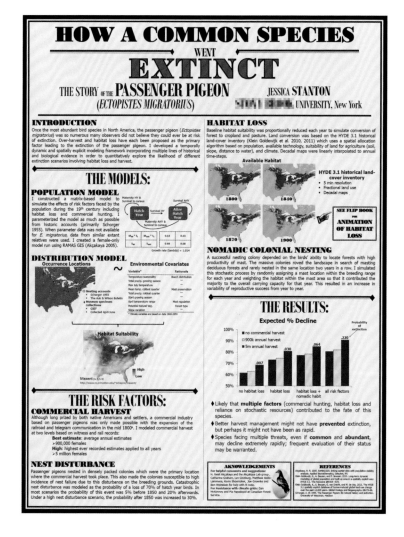

For one of my own works, I saw a typeface sample I liked. I wanted the warmth of the colors and the humanism of the typeface. So I stole the sentiment, as shown in FIGURE 7.14. Note that the poster is not a slavish duplicate of the image that inspired it. It turned out that the typeface was too limited for the task (no italic font), so I had to find another typeface that had some of the same qualities. The colors were modified to be closer to the colors in the photographs I was using, which happened to have

INSPIRATION

RESULT

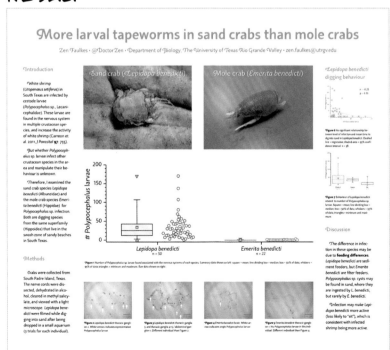

some of the same overtones as the font sample (the pink on the hand and the micrographs; grays that suggested blue in the animal pictures).

If you find something that you like, pick it apart and figure out what makes it work. Why does it appeal to you? Is it the layout? The color schemes? The typeface? The data visualization? Or some combination of those?

If you understand those design elements that someone else used, you can have some confidence that if you use some of those same elements, they will probably work.

Software

When people start to make posters, they often want to know what kind of software to use. You may be disappointed that I have not mentioned software much, and you may be further disappointed to learn that I refer to software as little as possible in this book, because this is not a book about software. It's a book about design.

If you understand the principles of graphic design, you can apply it to any poster, regardless of whether you draw it by hand or on professional graphic software. Many graphic professionals swear by the power of using hand-drawn sketches, or outlines scribbled on sticky notes, or other low-tech solutions, to develop their ideas and outlines (Seddon and Waterhouse 2009, Reynolds 2011). One of the advantages may be that when you start sketching, it is obvious that you do not have a finished product. Software is very good at providing a superficial polish early on that can convince you that you're further along in the design process than you are.

But most people will design posters using software. I will rarely refer to specific programs or apps, but I will often refer to tools that are common to many of them. "Many" is the operative word here, because most people will probably use several different software packages on a single poster. They have one package to make graphs, another to adjust photographs, another to write the text, and another to lay out the poster.

The core of poster design is graphic design, which is an area in which many standard office suites fall noticeably short. Be aware of the range of options, because there are big differences between graphic software packages. I've noticed time and again that many academics are – if I may be blunt – cheap. They don't want to pay for professional-level tools, and it often shows in the results.

At the one end of the simplicity scale is slideware, like Microsoft PowerPoint, Keynote, or OpenOffice Impress, which are usually installed on desktop computers by default and are seemingly everywhere. These are terrific for "quick and dirty" graphics, and many people (probably most people) make posters on these. But some of the techniques and options described in this book can't be done with these programs. They have quirks that often cause more harm than good. Some software works against good habits, and other software makes simple tasks complex.

At the other end of the scale are professional graphics packages, like Adobe Illustrator, CorelDRAW, or Inkscape. These are usually used to make single figures. Desktop publishing programs like Adobe InDesign, QuarkXPress, or Scribus are also options. These programs are created with multiple-page documents in mind, which is not needed for posters, but desktop publishing programs sometimes have more layout and printing options than graphics programs. These are feature-rich, powerful programs, but they take time to learn and are often more expensive. *Exception:* Inkscape and Scribus are open source and free.

There are only a few software packages in the middle zone of complexity. One that is surprisingly little known is Microsoft Publisher, intended for simple desktop publishing. It uses many of the same tools and has the same layout as the more popular PowerPoint, which makes it easy for PowerPoint users to learn. Unlike PowerPoint, it was designed with printing in mind, and has many layout features that PowerPoint does not.

But whatever software you use, invest the time to learn what it can and can't do.

Chapter recap

- The more abstract an image, the more work a viewer must do to interpret it.

- Graphic design is often about finding a balance between similarity and contrast. Too much similarity is boring. Too many contrasts are confusing.

- You can direct people's attention using cues such as size, position, and white space, which together create a visual hierarchy.

- Every poster should have an entry point.

- There is no best software for making posters. Understanding the principles of graphic design should let you make good posters in many different software applications.

CHAPTER 8

Figures

Deciding what to show deeply affects the kind of poster you make. If you have only text, you will be more limited than if you have charts and graphs. If you have only charts and graphs, you will be more limited than if you have illustrations and photographs. If the biggest point on your poster can be shown in one graph, you have more options than if you need to show many graphs.

This is also a good time to ask if you have elements that you have only ever written out in sentences and paragraphs that could be expressed visually. Dan Roam (2013) argues that most problems can be solved using pictures, which means most things can be represented as an image in some way. Remember, you only need six types of images to cover many situations (Roam 2013; Chapter 7).

Image types

Computers build graphics in two ways: pixels and vectors.

Pixel graphics (sometimes called bitmaps) create images from many small squares of different colors. This means that as you zoom in or enlarge, eventually you will hit the point where you start to see jagged edges. High pixel numbers are one of the desirable features that people look for when buying a digital camera.

Vector graphics describe shapes mathematically. Because the underlying equation for a square or circle or line never changes, as you increase the size, the image is displayed at the best resolution the printer or screen is capable of (FIGURE 8.1). But vector-based graphics are not very well known to most people, because there aren't many places where vector-based images are the norm. Cameras don't take vector-based images, and most images on websites aren't vector-based, either.

Pixel and vector graphics are so different that you probably need completely different graphics programs for each. For example, the company Adobe makes both Photoshop (a name now synonymous

ENLARGING PIXEL-BASED IMAGES

FIGURE 8.1

Comparison of pixel
and vector images
when enlarged.

ENLARGING VECTOR-BASED IMAGES

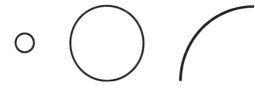

with image manipulation) for pixel graphics and Illustrator for vector
graphics. Similarly, Corel makes Photo-Paint for pixels and DRAW for
vectors. Currently, pixel graphics rule the roost, thanks to the web and
digital cameras. Search the internet for anything, and you will mostly find
images made from pixels unless you specifically look for vector graphics.
(SVG is the web's native vector graphics format.)

Pixel images

There are many different bitmap file formats, but the best known are JPG,
GIF, PNG, and TIF files. They are not perfectly interchangeable: each has
different limitations.

GIF images can only contain 256 colors, making them a poor format for
photographs or other images with lots of subtleties in color. But GIFs can
be animated, which makes for interesting prospects for digital posters or
posters created for social media (see *Digital posters* in Chapter 5).

The formats PNG, JPG, and TIF can show millions of colors, but none
allows animation.

Feature	GIF	PNG	JPG	TIF
Colors	256	Millions	Millions	Millions
Transparency	Yes	Yes	No	Yes
Animation	Yes	No	No	No
Compressed	No	No	Yes, lossy	Optional

TABLE 8.1

Comparison of pixel
image formats.

FIGURE 8.2

Transparency in different image formats.

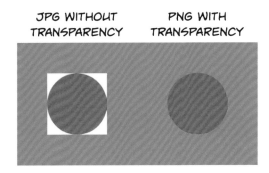

Images in JPG format are compressed to make file sizes small, which means the image is always degraded. Depending on how the JPG was saved, the amount of image fuzziness might be unnoticeable, but there is always a permanent loss.

Another key difference in pixel image formats is transparency. Every pixel must have a color in a JPG file, but pixels can be clear in GIF, PNG, and TIF files (FIGURE 8.2). The practical upshot of this is that if you have a pixel-based image that is not rectangular and you want to place it on a colored background, you will want to make sure your image is in a format that supports transparency. Otherwise, you will see a white rectangle around the shape you want.

The limitations of GIF and JPG formats make PNG files best suited for making posters. The TIF file format is not often used by graphic editors but is sometimes requested by publishers and printers (Chapter 22). Most graphics editors are able to convert between formats. Keep in mind that once an image has been converted down to fewer colors or a smaller number of pixels, it can't be converted back to a higher resolution.

Some of the file formatting details about the number of colors are mostly irrelevant for making paper posters, but these matter a lot for digital posters (Chapter 5), because the number of colors, compression, and animation impact file size. A poster made for social media might be limited to a 5 MB file size, for example. Under that limitation, you could make a GIF about 3,000 by 1,500 pixels. A PNG image would be much smaller, perhaps 1,800 by 900 pixels.

Image resolution is the other major consideration when working with pixel images. As noted above, as you enlarge any image made of pixels, sooner or later you will inevitably start to see the squares. You do not want that jagged look. You can always lower the resolution of a picture if necessary, but you can never increase the resolution of a picture (despite what you see on crime shows on television). To look good, you will need to know how many pixels an image has and roughly how big you want it to be full-sized on the poster.

The usual measure of pixel density is dots per inch (dpi). Yes, pixels technically are not "dots," because they are square, not round. And using old imperial measurements like "inch" is annoying for people living almost everywhere in the world (only Liberia, Myanmar, and the United States have mostly ignored the metric system). But thanks to American dominance in the desktop publishing industry, "dots per inch" is the phrase we're stuck with. A cheap magazine might have images printed at 100 dpi. Images from a high-quality professional printer, the sort used for a printed scientific journal, might be 1,200 dots per inch. The resolution sweet spot for posters is somewhere between 100 and 300 dpi. In almost every case, 300 dpi is more than enough (Campbell 2010).

Cameras often describe their image sizes in megapixels, which describes the total number of pixels in the image. But you want to know the number of pixels along the edge, so you know how many pixels will go into each inch of the final image on the poster. A camera might take an image that is 3,000 pixels wide. If you want to print that on a poster so that the final image is 10 inches wide, the final pixel density is 300 dpi – more than enough. If you print that image 24 inches wide, you have a pixel density of 125 dpi, which is probably fine, too.

Another issue is that because computer screens and paper have different resolutions, you can't figure out exactly how things will look on the page from a distance by making them small on a computer screen (Berkson 2010). A high-resolution image shrunk will be shown at the same screen resolution of about 100 dpi. A thin line can look thicker or thinner (if it's even visible at all) on a low-resolution screen than it will on a high-resolution printed page. This is particularly a problem with text, which often has thin lines in the letterforms. If you are putting together an oral presentation using slideware, you never have to worry about this, because you can see on the screen if the graphics are coming up short – but you do not have the luxury of immediately seeing a true "what you see is what you get" image when making a poster. When in doubt, print off at least a piece of your poster at full size and stick it up on a wall several feet from you to see how it will look.

Vector images

Vector-based images are awesome. A vector-based image is created using equations. The software says, "Draw a circle," or a line or a box or whatever, and then the hardware does the best job it can with that. This means that vector-based images scale to *any size* with no loss of resolution. Because they are described by geometry, vector-based images are fantastic for things like bar graphs that are made from lines, rectangles, and text. Common vector-based file formats are SVG, WMF, AI, and CDR. Some file formats commonly used for printing, including PDF and EPS files, are also vector-based.

If you're likely to be creating conference posters routinely, you should invest in a graphics package that can create and edit vector-based images. The results are worth it.

Distortion and resizing images

A common problem on posters is distorted imagery. This is usually a problem caused – again – by software making decisions for you by default. Sometimes, it happens because the software will try to "fit" an image into a predefined space. But more often, it's because of how the software resizes images.

Every graphics program allows you to resize images, but they may not keep the proportions of the images the same. If you grab a picture to expand it, a block that may have had a width-to-height ratio of 4:3 might become a block with a 4:2 ratio. People *will* notice the distortion.

FIGURE 8.3

Symbols and text distorted by careless resizing.

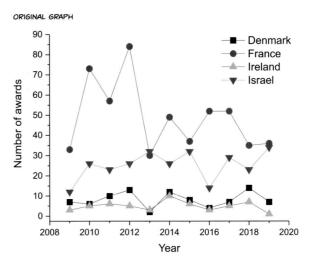

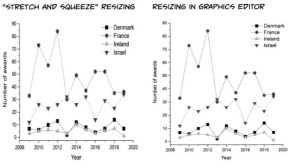

Graphics software will allow you to preserve the ratio when resizing in various ways. In some, there's a "lock ratio" or "lock aspect ratio" button or setting. Most also have a combination of keys and mouse buttons that will lock the aspect ratio on the fly. In PowerPoint, for example, pressing the shift key while resizing will keep the proportions the same. Resizing by grabbing corners in some software will preserve proportions, while grabbing the top, bottom or sides will not.

If you think you have made a mistake, some software will allow you to check the amount of enlargement, by showing "scale height" and "scale width."

If you need to make text bigger or smaller, change the point size or the font instead of dragging the text block. If you need to make a graph taller or wider, go back to the graph plotting software and change it there instead of resizing the graph in the layout program (FIGURE 8.3). More solutions are discussed in Chapter 19.

Image sources

There are many ways to get images for a poster, from "do-it-yourself" to using other people's work.

Photographs

The best graphic for a poster is often the picture you take yourself. Nobody else has seen pictures you've taken. Most people have high-resolution digital cameras available in their smartphones. Photography is a complex skill that is discussed in many resources, so I'll only mention a few suggestions here. Some tips are elements that we'll come back to in more depth later.

Look for third-party apps that extend your camera's capabilities. While many smartphones are feature-rich, there is no way that any phone can capture all the different ways that people want to use them. Features like time lapse, slow motion, and additional photo editing are more common in third-party apps than in the software bundled with your phone.

Wipe the lens. That thing's probably been in your purse or pocket for who knows how long. Get the crud off it.

Take photos in diffuse natural lighting if you can. If you are doing field work, a cloudy day can be better for pictures than a sunny one. Avoid using a flash, which creates intense light emerging from an unusual angle (horizontal from the phone, rather than from above). Smartphone cameras are very light-sensitive, so it is rare to need a flash. Watch for shadows that you cast, particularly if you are doing close-up work, or your subject is below you.

Take photos in the highest resolution possible. You can always scale images down, but you cannot "zoom in and enhance" like they do on crime shows. Likewise, if you have to crop a photograph later, a small section of a high-resolution photo can still be high-resolution enough for printing. Zooming in to an object using the phone's "digital zoom" is not a good idea, because most "zoom" functions on smartphones are done by software, not hardware. This means that "zooming" is done by making the pixels larger instead of getting more pixels. To get zoom without sacrificing the number of pixels, you need to buy a lens.

Most cameras have a "high dynamic range" (HDR) option that takes three pictures at different exposures, then stitches them together for a better image. The HDR setting can improve pictures significantly when the lighting varies across an image. It works best on still objects, since even a smartphone camera can't take three pictures simultaneously.

If you do have moving objects, try a "burst" setting (if available) to take several images, so you a better chance of a good one. Depending on your phone, you may be able take a high-definition video instead of a picture, and just grab the best frame.

Smartphone cameras are sensitive to shaking and blurring, particularly in HDR mode or low light. Find a solid spot to rest your phone or brace yourself on something solid.

Don't always center the subject of the photo. It is tempting to stick the horizon of a landscape right in the middle of the frame, or a vertical object cutting the photo into left and right. But these compositions can be boring. Photographers often talk about the "rule of thirds," where they try to place the most interesting part of the picture at a third or two-thirds across the photo. Many cameras have a grid option that divides the picture into thirds. For similar reasons, diagonal elements often add interest to a photo.

Take pictures from perspectives other than what you see at eye level when standing. Low angles can make people and objects look more impressive. Smartphone cameras often have very short focal distances, so you can get in close to very small objects, particularly if you use manual focus.

Leave room on your photo for overlaying text later. Because you are taking this picture to be part of a poster, you may want to label the image. A boring part of the photo with a consistent color and minimal detail is an excellent place to put text directly on top of the photo, rather than forcing copy off the picture.

Internet images

Searching the internet for images is so easy that its most people's first stop to get images for their poster. The ease of searching is balanced out by several pitfalls.

Many people pull images from the internet to use on their posters and don't check the number of pixels. Many people make the mistake of taking a low-resolution bitmap image from a website, often an institutional logo, without realizing that the image is only a few hundred pixels wide. They then put it on a poster and print it at several hundred times the size it was intended to be displayed at. The resulting image looks horrible.

For example, if you have an image that is 800 pixels wide, and you want to print it 24 inches (610 mm) wide, the print resolution will be about 33 dpi, and that will look like rubbish. It'll still look poor if printed at 12 inches (305 mm) wide (800 / 12 = 67 dpi).

"Found on the internet" does not mean "free to use." Unless an image is explicitly tagged as "public domain," it may still be copyrighted (Seddon and Waterhouse 2009) or have its use restricted by licenses. "Public domain" images are not protected by any sort of intellectual property laws and cannot be owned. Anyone can use them in any way at any time for free. In contrast, "copyright" means that someone (whether an individual, corporation, or institution) has the exclusive right to make copies of that thing, adapt it, or distribute it. In most jurisdictions, copyright is automatic. That is, no special steps need to be taken to register that a creator has copyright over some content.

Whether it is okay to use some image is difficult to say, because copyright laws vary from place to place and can be notoriously arcane. Science fiction humorist Douglas Adams (1985) wrote:

> If you wish to understand [copyright], and I would suggest to you that you do not, then you should first limber your brain up with a little gentle quantum Mechanics, work out on Relativity, try really pumping lobe iron on a Unified Field Theory and then take a short cooling rest. Now you should be ready to think about copyright law …

Another point of confusion is that some images are described as "royalty free." Again, this does not mean "free to use." Royalties are payments for repeated use of something: the more uses, the more royalty payments. A "royalty free" image usually requires a "one and done" payment for use.

There are many ways that content creators can give explicit permission that it is okay to reuse something. The concept of "reuse as you want" is sometimes called "copyleft," as in the opposite of copyright. A popular way to give explicit permission to reuse content is by using Creative

Commons licenses. Some online image resources allow you to filter for images that have explicit permission to use. Look for "public domain" or "Creative Commons licenses" filters.

There are six different kinds of Creative Commons licenses that have gone through several revisions (Creative Commons 2019). Some Creative Commons licenses are more relaxed, and some are more restrictive. Moreover, the legal interpretation of what is permitted under a license is sometimes non-intuitive, particularly for terms like "non-commercial" (Moody 2014, Butler 2015, Drauglis v. Kappa Map Group 2015).

It is unlikely that anyone will get into legal trouble for having any image on an academic poster, regardless of license or copyright. Your poster is a one-off, being shown to a limited audience over a short time, and you're not going to sell the poster. But if you upload a digital copy of your poster to an archive or file-sharing site online, you increase the risk of a creator saying "Please take that down" or worse.

Regardless of the low chance of facing either legal action or social embarrassment from a creator, respecting copyright and licenses for material is the right thing to do. Creating high-quality images is hard work, and people should be able to benefit from that work if they choose.

Stock photography

There are many stock photography services that have large libraries of high-resolution images that come with copyright clearance for a reasonable fee. These have the advantage of being professionally shot. There is often a surprising range of images on sometimes obscure subjects.

If you decide to buy a stock photo, check the resolution. Many companies scale the price of the photo to its resolution: higher resolution means a bigger price tag. You may not need the highest-quality image, but you do need to do the calculations of its final size.

A downside is that the images are not new. Stock photo catalogues can contain old and dated materials. If you are trying to show that you are doing current research, you may not want an image featuring someone with a hairstyle that went out of fashion decades ago. People may have seen those exact images before in a different context, and this simply looks lazy. Many businesses have been mocked for showing stock photos instead of real employees, while claiming to value diversity.

But even if people haven't seen the exact images, they have probably seen clichéd images very much like them (Douglas 2016). Women in stock photos wear white clothes, do yoga, and eat salads alone while laughing more often than the general population (Zimmerman 2011, West 2012). Searching for stock photos of scientists brings up pictures

FIGURE 8.4

Stock photograph
of biologist.

of people in personal protective equipment staring at colored fluids in glassware (Becker 2018).

Sometimes the images in stock photos are ridiculous. The photo of a biologist in FIGURE 8.4 raises questions. What is a plant doing so quickly that it needs to be timed? Why can this lab not afford a digital stopwatch instead of having to use someone's egg timer?

The worst use of a stock photo is to print a watermarked preview image taken from the web. It marks you as incompetent because you didn't recognize the watermark, lazy because you didn't care about the watermark, cheap because you wouldn't pay for the full version, or some combination of the three.

Chapter recap

- Posters are a visual medium, and graphics provide the fastest, best way to inform viewers.

- Because posters are a large format, care needs to be taken to make sure images are not distorted or pixelated.

- Take your own photos when possible.

- Respect copyright and licensing restrictions.

CHAPTER 9

Presenting data

For most academic projects, the data are the heart of the whole thing. Most data need to be presented in a form that is quick to grasp and more precise than a verbal description. You will usually need to show data visually using graphs. There are usually many ways to present the same data, and those decisions can make a huge difference to how you design the poster.

Organizing data

Regardless of the form you use to present information – text, graph, or table – it helps for the data be organized in consistent ways. There are five strong ways to structure information to help make it meaningful (Wurman 2001):

Location – For physical objects, location is often the most relevant feature. When you get in an elevator, you expect buttons to be arranged from the top floor to the bottom floor. If you search for "gas stations," you are probably low on gas and want to know the closest one. You may want to arrange cites from east to west, or north to south. Location may not be a literal physical location in space, however.

Alphabetical – While the order of letters in an alphabet is arbitrary, it is so familiar to native speakers of a language that it can be used to structure information that might otherwise not have an obvious structure. If you want to list famous movies of 1933, listing them alphabetically allows people to locate individual entries quickly.

Time – Chronological order is useful for understanding and highlighting sequence. Most historical events lend themselves to this sort of structure. For instance, movies might be ordered by release date.

Category – Bookstores, libraries, and video services use categories to order information, only they call them "genres." Someone who wants to read a murder mystery probably appreciates all the detective fiction being together, not interspersed with romances or business manuals. Color can also be a powerful organizing category.

Hierarchy – This can be anything that can be scaled along a continuum. This is a powerful way to organize numeric data, but also lends itself to business organization structure. Colors and grayscales are another example of continuous variables that can be arranged hierarchically.

The LATCH mnemonic can help you remember these options.

These sorting strategies are not mutually exclusive. We might sort by category first ("bakery"), then time ("what's open?"), then location ("what's closest?"). It's also worth noting that if you have two related variables, you can organize the data by either of those variables. For example, if you had a list of all the episodes of a television series and their viewing ratings, you could organize the data by the chronological date of the episodes or by ranking episodes from best to worst.

Graphs for analysis

While you may not end up with graphs of all your data – some might be better shown some other way – you should always plot your data as part of the analysis. Not only are images more easily understood (see FIGURE 7.1; McCloud 1993, 2018), they also let you make new insights much more easily than if you are just comparing statistical test results (Few 2009, Roam 2013).

TABLE 9.1 shows four sets of paired data called "Anscombe's quartet" (Anscombe 1973). They might be the independent and dependent variables from replicated experiments.

A		B		C		D	
x	y	x	y	x	y	x	y
10.0	8.04	10.0	9.14	10.0	7.46	8.0	6.58
8.0	6.95	8.0	8.14	8.0	6.77	8.0	5.76
13.0	7.58	13.0	8.74	13.0	12.74	8.0	7.71
9.0	8.81	9.0	8.77	9.0	7.11	8.0	8.84
11.0	8.33	11.0	9.26	11.0	7.81	8.0	8.47
14.0	9.96	14.0	8.10	14.0	8.84	8.0	7.04
6.0	7.24	6.0	6.13	6.0	6.08	8.0	5.25
4.0	4.26	4.0	3.10	4.0	5.39	19.0	12.50
12.0	10.84	12.0	9.13	12.0	8.15	8.0	5.56
7.0	4.82	7.0	7.26	7.0	6.42	8.0	7.91
5.0	5.68	5.0	4.74	5.0	5.73	8.0	6.89

TABLE 9.1

Data from Anscombe's quartet.

All four sets have the same summary statistics. The means and standard deviations for x are the same (9 and 3.31662, respectively). The means and standard deviations for y are almost the same (to two decimal places: 7.50 and 2.03, respectively). The correlation between x and y, measured by Pearson's r, is the same (to three decimal places: 0.816). The slope of the line described by a linear regression is the same to a couple of decimal places ($y = 3.00 + 0.500x$).

It difficult to tell quickly from the table that the data are different. But a plot makes it instantly obvious not only that they are different but how they are different (FIGURE 9.1).

While the summary statistics might not indicate any difference in these data sets, the interpretation of these plots should be completely different. Plot A shows a messy correlation. Plot B shows a precise non-linear relationship. Plots C and D show clear outliers.

This quartet of graphs was almost certainly created using a trial-and-error method to make the summary stats the same. But with the aid of computers, the point can be made more emphatically (Matejka and Fitzmaurice 2017). Data sets with the same summary statistics (to two decimal places) can be created that will reveal a dinosaur (nicknamed *Datasaurus*), a scatter, a bullseye, or a star when plotted (FIGURE 9.2). There are nine more variants not shown here.

FIGURE 9.1

Anscombe's quartet.

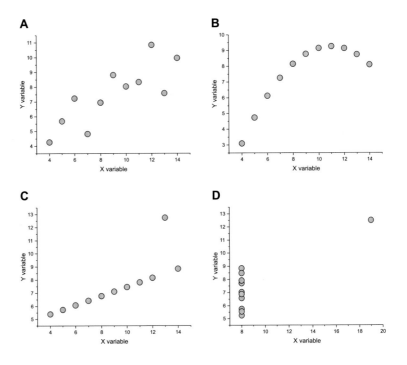

BETTER POSTERS

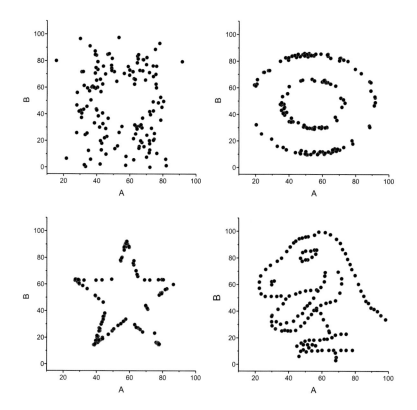

FIGURE 9.2

Datasaurus and three of a dozen graphs that have the same summary statistics as *Datasaurus*.

Some people have used these examples to argue that showing summary statistics alone (e.g., averages in a bar graph) is poor practice (Weissgerber *et al.* 2015), suggesting that using these common summary graphs shows either professional incompetence or deliberate malfeasance. The extreme position is that every graph should contain all raw data points. But this misses the difference between data exploration and data visualization. The point of both summary statistics and visualizations is to simplify and condense (Fry 2019), and there will always be a trade-off between the need for something that is comprehensible at a glance and the possibility of missing an interesting pattern.

It is wrong to say that bar graphs always misrepresent the underlying data. The creators of Anscombe's quartet (Anscombe 1973) and the dozen variants of *Datasaurus* (Matejka and Fitzmaurice 2017) labored mightily to get such different data sets generating the same summary statistics. If you examine your data and find that the data sets have similar distributions and skews, and no outliers, a bar graph can represent the data truthfully and quickly. Choosing an unfamiliar graph instead of a familiar one gains you little and costs more (Roam 2013, Duarte 2019a).

Graphs

Once you have decided how to organize the data, decide how to show the data. The type of data will often suggest, if not dictate, the kind of graphs you have. Remember that there are six basic ways to show something: a portrait, a chart, a map, a timeline, a flowchart, and a multi-variable plot (Chapter 7; Roam 2013). (Note the overlap between these and the LATCH organizing principles.)

You may have used complex graphs for analysis, but you should look closely at whether you need these graphs for presentation. Nancy Duarte (2019a) advised:

> Using charts that are more complex than they need to be adds mental labor to the reviewer and pulls attention away from the key insight … Complex charts can be fascinating and look impressive, but they often obscure the main point. Everyone processes and understands the bar chart (measures quantity), pie chart (measures parts of 100 percent) and the line chart (measures changes over time). Everyone understands these charts. **Everyone!**

Dan Roam compares the process of selecting graphs to selecting a meal. In a meeting situation, if you need to feed people, you get pizza. Pizza is simple to order, easy to eat, everyone knows it and likes it. You don't bring an exotic gourmet meal to a lunch meeting, because it is hard to order, difficult to eat, needs a lot of explanation, and may need a lot of time to get people even to try it (Roam 2013). Most graphs on a poster should be like pizza, not a gourmet meal.

Graphics practices change over time, however. One increasingly popular suggestion is report effect sizes rather than p values in graphs that include statistical analyses (Ho *et al.* 2019).

General principles

More than almost any other element on a poster, graphs are prone to Roam's observation (2013): "Computers make it too easy to draw the wrong thing." Many common software apps used to create graphs make amateurish mistakes in plotting data, as shown in FIGURE 9.3.

There are many kinds of plots and visualizations for data besides the common ones described below. But regardless of the type of graph used, the guidelines below can apply to all of them.

Don't recycle graphs you made for some other purpose. It's likely that your old graphs have the wrong proportions for the layout of your poster,

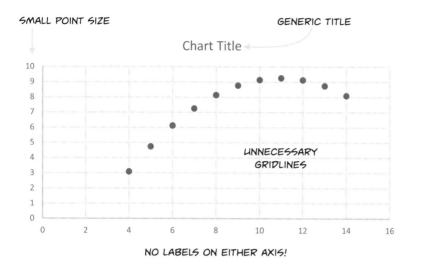

SMALL POINT SIZE

GENERIC TITLE

Chart Title

UNNECESSARY
GRIDLINES

NO LABELS ON EITHER AXIS!

FIGURE 9.3

Graph created using
software defaults.

are too detailed, and use the wrong colors. Make graphs specifically for your poster.

The most common reason to go back to the original graphics program is to resize graphs. Graphs generally look better when they are wider than tall (Tufte 2001), but different graphics programs and displays have different default height-to-width ratios (FIGURE 9.4). As you're laying out your poster, you will probably find that sometimes things fit more readily if your graph is a little taller or a little narrower. Do not stretch the graph (see *Distortion and resizing images* in Chapter 8). Distortions look horrible. Always go back to the original graphics file, change the proportions, and export the revised graph.

Speaking of ratios, people will sometimes mention the "golden rectangle" to me in discussing poster design and extol its virtues of perfection. In a "golden rectangle," the wide edge is about 1.618 times longer than the narrow edge. (I say "about" because the actual proportion, like pi, is an irrational number with an infinitely long set of digits.) These two

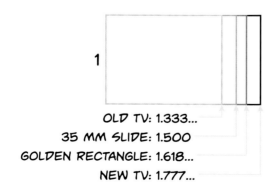

1

OLD TV: 1.333...
35 MM SLIDE: 1.500
GOLDEN RECTANGLE: 1.618...
NEW TV: 1.777...

FIGURE 9.4

Common
display ratios.

FIGURE 9.5

Chartjunk.

THIS IS A **COMPLICATED** WAY OF SHOWING TWO NUMBERS!

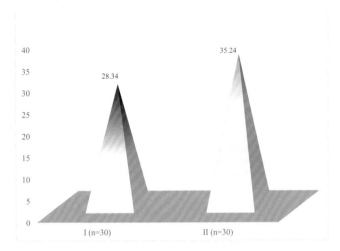

measures form the "golden ratio," which has recursive properties. But I have never found the "golden rectangle" useful in design. That the sides of the rectangle are not proportions that can be expressed in simple integers is frustrating and makes creating evenly spaced grids needlessly difficult. Much of the talk about a golden rectangle being an idealized shape is kind of unproveable rubbish.

Remove chartjunk (Tufte 2001). Chartjunk is anything that is irrelevant to, or obscures, the information the graph is trying to show. Showing bars in a graph in a 3-D perspective is a common example of chartjunk (FIGURE 9.5).

This isn't just an aesthetic consideration. Chartjunk can obscure the data or mislead you about the data (FIGURE 9.6).

FIGURE 9.6

Ribbon graph.

WHAT'S THE ORDER OF THE RIBBONS?

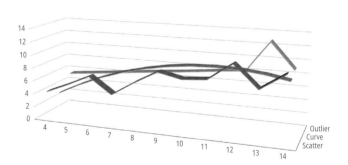

The nicest thing that can be said about 3-D bar graphs is that some people like them. They don't always impair accuracy of judgment, although making a judgment does take longer than with a simpler 2-D graph (Carswell *et al.* 1991).

Many researchers' instinct is to democratize data: show them all and treat them all equally. This approach might have merits in a print medium where someone has the time to explore deeply, but for a poster, don't just show the data. Use contrast to show what is important about the data. For example, you might use a color to highlight the most common response in a survey, or a group that is statistically significantly different than the others. If you want people to focus on the median in a boxplot, emphasize it by making the median line a heavier weight or different color than the box. Use contrast to highlight trends and differences in your graphs, or elements that you want to bring attention to.

A good trick with graphs is to keep most of the data in a neutral color (e.g., gray). Use a color to highlight key points or a trendline (Jambor 2019, Murray 2019). For example, if you are comparing three groups in a bar graph and one is statistically different from the others, you might change the color of the bar for that group.

Remember that similar colors imply similarity between groups. Unconnected categorical data should be shown with unordered colors, rather than several shades of the same color (Stefaner 2019).

While creating your graphs, work on captions and legends that will go with them. Figure legends are probably more important than the main text. In magazines, people read photo captions more than the body text of the article (Ogilvy 1963) and the same is probably true of posters. If you are going to use a picture, you must be ready to talk about it. Creating a graph is a very different skill than explaining that graph. "All good pictures do *not* need to be self-explanatory, but they do need to be *explainable*" (Roam 2013).

Avoid abbreviations in your axes or legends. People often run statistical analyses using short "nicknames" for variables, and these nicknames end up in the graph, sometimes with no explanation of what each means in the main text.

Portraits

Portraits are depictions of objects and other situations where you need a qualitative representation rather than a quantitative one. "Portrait" is being used here in Roam's sense of a generic representation of a person or thing (Roam 2013). For example, a biologist discovering a new species will want to show what it looks like. A chemist making a new chemical

FIGURE 9.7

Graphic using
human for scale.
(Remes *et al.* 2009)

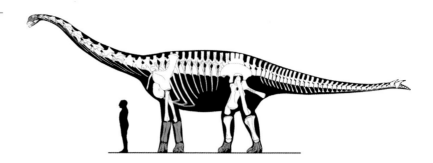

compound might want to show its chemical structure at the molecular level and its consistency and color at the macro level.

When you are showing something large, small, or unfamiliar, include a scale. Micrographs of cells should have scale bars. Large or unfamiliar objects, such as extinct animals, are often scaled by including a silhouette of a human (**FIGURE 9.7**; Remes *et al.* 2009). If the representation is not to scale, say that, too.

Maps

Maps represent location. Sometimes location isn't shown by a literal cartographic map. A picture of a place can often represent a physical location. "Where" can also be more abstract than a physical location. For example, an org chart is a map of employment hierarchy. A Venn diagram is a map of common properties. A pedigree, a family tree, or a phylogeny is a map of relationships. Ideas can be related by a concept map. We use the word "map" for geography, but anytime you use terms like "above" or "below," or "closer" or "further," you have potential to create a map-like graphic.

There are now many websites and software packages that can generate maps (Zastrow 2015), but perhaps more than most graphics, mapmaking is a human skill (Stevenson 2012). There are too many places with names for software to handle placement of those names on a map. An award-winning map of the United States created by cartographer David Imus required thousands of hours to complete because there was no substitute for placing things by eye (Stevenson 2012).

This level of devotion to craft may not be necessary for a poster graphic, but it is a reminder of the limitations of software tools. As in so many other cases, popular geographic websites and mapmaker programs are not necessarily the best for posters.

Here are some issues to consider if you plan on making a traditional map showing geography.

First, some online geographic services do not export maps in standard graphic forms. Map data may be saved in SHP and KML files (GISGeography 2020), not PNG and WMF files. Many people take screenshots of their computer display to circumvent this. But this creates images with a limited number of pixels, and might require stitching together multiple images in a graphics program to create one large enough for a poster.

Second, online map services may watermark their maps with logos that need to be removed.

Third, if you make a geographic map, determine the purpose of your map. This deeply affects the choices of display. For instance, a geographic map might show water bodies, elevation, soil types. A political map might show cities, roadways, and borders of counties, parishes, and municipalities. If you need to make more points, use more maps rather than trying to create a single map containing complex overlaid data.

Fourth, provide a context for your map. People's geographic knowledge is often patchy. Taking a map of the world into a public place, asking people to point out countries, then watching them fail miserably, has been fodder for television comedy for years. If you are at an international conference, people might not know the location of your country on a globe. If you are at a national conference, people might not know the location of your province or state. Provide a small inset that shows where a location fits within a larger region people are more likely to know.

Fifth, provide a scale for your map. A scale is one of the most forgotten elements on a map. You cannot assume that people know how big the place in the map is.

Sixth, maps should attribute their data sources – both the geographic data for the map and any other information that is plotted there. Ideally, these should be on the map, but a figure caption might work well too.

Finally, put a border (neatline) around your map. This doesn't affect the data, but it is part of mapmaking tradition and makes it look nice. When maps were crafted by hand using ink, neatlines used to be ornately decorated, but a thin line works better now (Freitag 2012).

Bar graphs

Amounts are usually shown in bar graphs. Bar graphs are among the most common chart types, so they are familiar and simple to interpret. Most bar graphs are vertically oriented. One paper suggested that this is because we are familiar with piles getting taller as we stack more objects on top of them. But horizontal bar graphs are sometimes useful, particularly if labels are long (FIGURE 9.8).

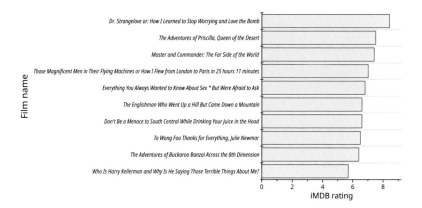

FIGURE 9.8

Horizontal
bar graph.

Categorical data can be shown with lines instead of boxes, but viewers strongly associate line graphs with trends. They tend to describe a line graph using phrasing like, "As *x* increases, *y* ..." (Zacks and Tversky 1999, Shah and Hoeffner 2002). One viewer described a line graph showing height in men and women (shown as discrete categories) by, "The more male a person is, the taller he/she is." This description was probably prompted by the format of the graph, rather than by a recognition that gender falls along a continuum.

Bar graphs often show averages, not raw scores. If you are analyzing data with a t-test or analysis of variance (ANOVA), you are comparing averages, and this should be clear to a viewer. Error bars are good cues to show that you are plotting an average instead of a single quantity. Error bars can represent many different things: standard deviation, standard error, confidence interval, or minimum and maximum – and there is no standard way to distinguish these (Cumming *et al.* 2007). The most common problem with graphs on posters is not saying what the error bars show. You must tell your audience, because the interpretation differs. The error bar descriptor can be placed in a figure legend, but I prefer to put a label in the y-axis (e.g., "+ S.D." for standard deviation). This is a good general principle in graphic design: **put information at the point of need**.

Box plots (also known as box and whisker plots or range bars) show data with more complexity (Spear 1952). They are intended to show averages and how the data are distributed. Many box plots show five values: median, minimum, maximum, and the upper and lower limits of 50% of the data (quartiles). But box plots are not standardized in their display. Some versions of box plots show means and/or outliers. Other use the whiskers to show 95% of the data, confidence internals, or something else. As with error bars on a simple bar graph, box plots need some sort of description of what each part of the plot shows, particularly the whiskers.

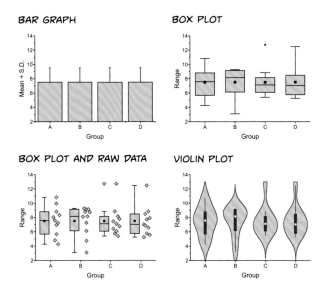

FIGURE 9.9

Four ways
of plotting
quantitative
categorial data.

As Anscombe's quartet and the *Datasaurus* imitators showed, summaries can be deceiving (even with measures of variation like standard deviation). Therefore, many people prefer even more complex types of plots which show some of the underlying structure of the data (FIGURE 9.9). Showing raw data is useful when the sample size is small. Violin plots can distinguish different data sets that are identical by summary statistics or even box plots (Matejka and Fitzmaurice 2017).

Be thoughtful about the x- and y-axes. Darrell Huff (1954) showed how to make misleading graphs by selecting where to start an axis. The trick is so well known that "Start every axis from zero!" is both common advice (Few 2009) and a stock criticism (Yanofsky 2015). But Huff never explicitly suggested "Start every axis at zero" as a rule. There are many cases where starting at zero is misleading (Yanofsky 2015, Sauro 2016). For example, there is no reason to start a graph of human body temperature at zero in any temperature scale (FIGURE 9.10). Zero is not part of the range that is relevant for most purposes. Furthermore, focusing on zeroing axes overlooks other misleading games people can play with axes, like adding breaks (Huff 1954).

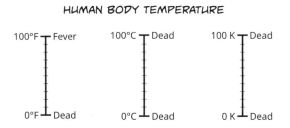

FIGURE 9.10

Temperature ranges
in the context
of relevance to
human bodies.

Percentages

Percentages, proportions, and fractions are simple numbers. But for such simple data, there sure are a lot of ways of showing them. You can use pie charts, donut charts, stacked bar graphs, separated bar graphs, and more.

Pie charts are divisive (Roam 2013). They are loved by many but criticized by others as inaccurate and dumb (Tufte 2001, Few 2007).

Pie charts quickly convey that they are showing a relationship between parts and a whole. Indicating the type of data quickly is a huge advantage on a poster. People can quickly grasp what fraction of a whole is shown, more so than with stacked bars (Simkin and Hastie 1987, Shah and Hoeffner 2002). But pie charts fail beyond very small data sets. As the number of categories increases, the size of each slice gets smaller. Small proportions are difficult to see and label (Few 2007). Whereas a bar graph can be rescaled so the height of the bars fills the page, a pie chart cannot. It will always be a circle.

Pie charts are okay if you ask people to make very simple comparisons, such as when there are two or three categories of quite different sizes. But people are bad at comparing angles (Cleveland and McGill 1985, Simkin and Hastie 1987), particularly when none of the edges of the angles are aligned. This makes pie charts a poor choice if the purpose of the display is to compare percentages that differ by just a few points. Hence, many people will label the slices with the value they represent, which defeats the purpose of the graph in the first place (Few 2007).

FIGURE 9.11

Rotated pie chart.

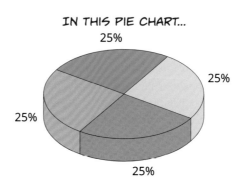

IN THIS PIE CHART...

25%

25%

25%

25%

THESE ANGLES REPRESENT THE *SAME* NUMBER!

Rotated pie charts are even worse (FIGURE 9.11). There is not even a pretense of the angles corresponding to the units displayed. Unsurprisingly, viewers are worse at making comparisons with 3-D charts (Siegrist 1996).

If you do use a pie chart, make sure the edge of one slice is aligned to the horizontal or the vertical (Simkin and Hastie 1987). This helps "anchor" the graph to cues that are relevant and familiar.

Donut charts are strange, because they require a viewer to compare arcs, which is not something people are often asked to do. However, it seems that it might be easier than comparing angles because arcs are lines, and we are better at judging line length than angles. But while you can label the slice in a pie chart directly if the slices are large enough, you can rarely do this for slices in a donut. Further, there's the question of what to do with all the empty space in the center of the donut.

Stacked bar charts can be a little better at showing percentages, particularly when you compare percentages of several different groups. Stacked bar charts are still not ideal when you have three or more categories. People are much better at judging length, but this becomes difficult when the start or end of the lines are not aligned along a clear horizontal or vertical axis. In a stacked bar chart, the boxes at the top and bottom will align with the x-axis, but it's unlikely the boxes in the middle will do so.

Separated bar graphs have the advantage of being the easiest to compare, but they can become complex to read if you have multiple groups.

If you have very few percentages, sometimes the easiest and fastest thing to do is just to show the numbers, as large as possible. An example of a poster using large numbers instead of charts is shown in FIGURE 9.12.

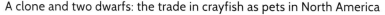

A clone and two dwarfs: the trade in crayfish as pets in North America

Zen Faulkes, Department of Biology, The University of Texas Rio Grande Valley

FIGURE 9.12

Poster showing percentages using numbers.

Time series

Time series and other continuous variables are commonly shown as line graphs. That time is so often a variable of interest (past, present, future) makes "change over time" one of the most widely used plots. Generally, time will be shown on the x-axis. Graphs that break this pattern are difficult to interpret (FIGURE 9.13; Jambor 2019).

While line graphs imply that data are continuous in theory, the data need not be continuous in practice. For example, "sampling over a year" may produce 12 data points if samples are taken monthly or 365 data points if it is done daily. When there are few data points along the x-axis, you might want to show a line plus a symbol to indicate the interval between the points more clearly. For many data points, the symbols become impossible to resolve and you should just use a line.

Straight lines are usually best for connecting data points. Some software programs offer the option of connecting points with a "smoothed" curvy line. This can wrongly imply that data were sampled continuously when they were not. The curves sometime can also dip above or below maximum or minimum recorded values (FIGURE 9.14). If you choose to use a smooth curve, use symbols to clearly indicate the actual plotted points, to show that the curve is an extrapolation.

Relatively thin lines can look good on a line graph, but the thickness of the line can be difficult to judge on a computer screen. In preview mode, a hairline thickness may look the same as 1 point, but these will become obviously different when printed. Because graphs are viewed from a distance, it is worth printing out a test copy of line widths and checking them from a distance.

Sometimes a good option for graphs is to fill the space underneath them, turning a line into a more visible shape (FIGURE 9.15).

The skill with line graphs comes when there are multiple lines. The point of line graphs is that each line must be able to be followed, so if lines cross over each other, you must be able to connect a line all the way along its

FIGURE 9.13

Line graph with time on the x-axis as expected (left) and with time on the y-axis, the opposite of expectations (right).

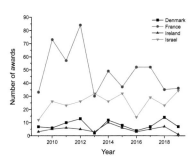
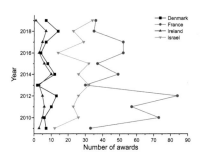

LINE SMOOTHING CAN BE MISLEADING!

STRAIGHT LINES

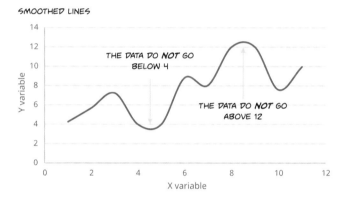

SMOOTHED LINES

THE DATA DO **NOT** GO BELOW 4

THE DATA DO **NOT** GO ABOVE 12

FIGURE 9.14

Smoothing in line graphs.

LINES ARE HARD TO SEE FROM A DISTANCE. FILLED SHAPES ARE MORE VISIBLE!

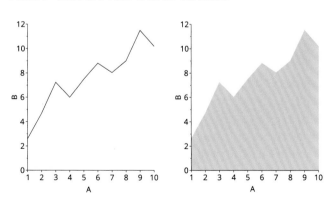

FIGURE 9.15

Line graph compared to area graph.

length. If you are using symbols, use different symbols for each line. Use different colors or dash patterns for each line. Keep in mind that each additional line creates more visual confusion and can make the graph hard to interpret. In some cases, small multiples may be better. Stacked line graphs are better than side-by-side ones (Few 2009) because the x-axis is usually the same. This can make comparisons for, say, a set length of time easier.

As with bar graphs, line graphs are sensitive to choices about the axes, particularly the y-axis. If the y-axis is too compressed, trendlines become flat, giving the impression of no change. If the y-axis is too extended, small changes emerge like mountain ranges, giving the impression of chaos. If possible the average slopes in a line graph should be about 45 degrees (Tufte 2006, Few 2009).

Some line graphs show averages and, as with bar graphs, should have error bars. If there are multiple lines on the graph, the upward and downward error bars may collide. Because error bars are symmetrical, you can remove the upper or lower one and still be accurate.

If you have multiple lines, label them directly if possible rather than using a legend (Shah and Hoeffner 2002). A legend forces you to go back and forth.

Scatterplots

Correlations are usually shown as scatterplots. Scatterplots contain data points, and sometimes a trendline or curve fit. Scatterplots are complex, so you should always be asking what the point of the chart is and whether it can be replaced by a simpler data display (Roam 2013). Because scatterplots imply relationships, ask yourself whether the relationship might be spurious or caused by some common factor not shown in the graph (Roam 2013). If you don't ask this question, chances are good that someone visiting your poster will.

Symbols for data points in a scatterplot can vary by exterior shape, interior markers, color, and size. Use symbols that can distinguish a single data point from several data points close together. Squares and diamonds can be hard to resolve because their straight edges can line up. This can't happen with circles or triangles unless they are exactly on top of each other, although when two points are very close, it is hard to resolve them no matter what shape they are.

An even better approach is to outline the points with a contrasting color so that edges are visible (FIGURE 9.16). Changing the contrast of the outline is an effective way to change the visual weight of the individual data points.

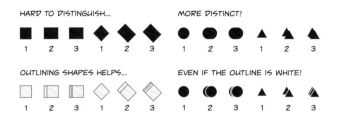

DATA SYMBOLS IN SCATTERPLOTS

HARD TO DISTINGUISH...   MORE DISTINCT!

OUTLINING SHAPES HELPS...   EVEN IF THE OUTLINE IS WHITE!

FIGURE 9.16

Legibility of symbols
in scatterplots.

When there are multiple symbol types in a scatterplot, they should differ in two ways, not just one. That is, two data sets should not be blue circles and orange circles if they can be blue circles and orange triangles.

The best size for individual points will vary with the number of points in the plot, and their position within the graph. When there are few data points, larger symbols ensure viewers can see them. But as more points are added, they can overlap. If the overlap is too great, the graph becomes a barely interpretable blob. Small points can be visible if there are enough of them close together.

The size of data points should not indicate different types of data. Some software can create "bubble plots" where the size of the point indicates how many data points are represented. This is sometimes useful for integer data. For example, if you count the number of individuals every day of the week, there are only seven possibilities for day of the week and the number of individuals will always be an integer. If the number of individuals is consistent, there could be many points occupying the same location on a graph.

Trendlines are often added to scatterplots to show predicted relationships between variables. Trendlines are incredibly valuable on posters because they can be interpreted more quickly than the individual data points. It is often valuable to give more visual weight to the trendline by increasing its contrast so it can be seen quickly from a distance (see FIGURE 10.4). The individual data points can be shown with lower contrast so they are visible to those who want to take the time to examine the graph in more detail.

The relationship between a trendline and data points in a scatterplot is a good example of the need to orient viewers in a graph. When people are performing a visual task, they go through predictable stages. They orient themselves first and use pattern perception to look for major differences and trends. They are looking for whether a line goes up, goes down, or stays flat; and for whether it is smooth or wave-like. They are not looking at the details of the axes. (Edward Tufte [2006] exploited this with "sparklines": word-like graphs showing trends over time, with no axes

FIGURE 9.17

"Convincing," from
the xkcd webcomic.
(Randall Munroe,
https://xkcd.
com/833)

or labels.) Only after they know "which way is up" do they inspect the
details in the graph (Roam 2013) – if they are interested enough. This is
one reason why standard graphs perform better than more exotic forms:
people can quickly tell "which way is up" and they know how to orient
themselves so they can spend more time on the details. Non-standard
elements make people spend longer orienting themselves.

In a limited time situation like a poster presentation, people may not
have time to dig into the data on a graph. This means you want to make
a trendline obvious, not the data – but you still want the data to be there.

Make sure your axes are labeled, with units in the axis (FIGURE 9.17;
Munroe 2010). Remember, not every piece of software adds axis labels
automatically (FIGURE 9.3). And while you're looking at your axes, see if
you can simplify them by using fewer tickmarks and labels.

Tables

Tables are probably the worst way to convey information on a conference
poster. A table requires more reading and interpretation than a graph.
Consequently, people take longer read a table than a graph, and novices
interpret the information less accurately when faced with a table (Cook
and Teo 2011). Tables can be precise, which is great when someone needs
a lot of detail and has a long time to study the numbers. But this does not
describe the typical conference goer.

Tables should be replaced by graphs whenever possible – and it usually
is possible. Almost any information that can be placed in a table can be
shown in a graph (Gelman *et al.* 2002). Tables are particularly suscep-
tible to bad choices imposed by software by default. When you are first
creating a table, you need to see where the cells are, so by default there are
a lot of lines to show where you can put data. But you have to take them
away, because the software doesn't (Butterick 2016). Tables in spreadsheet
software are optimized for data entry and manipulation, and tables in
graphics programs are an afterthought. The difference between tables in

journals, typeset by professionals, and most tables on conference posters is profound.

Tables often fall victim to clutter. Most people have tables with heavy grid lines (FIGURE 9.18), which Tufte (1990) called "data imprisonment." Others have more stripes than a zebra to show the columns, the rows, or both.

The simplest thing that will improve table is to remove all the vertical lines and stripes. White space organizes the cells better than lines can. It's usually a good idea to remove most of the horizontal stripes and lines, too. If you do use any lines, make them thin. Many tables have lines dividing the header from the data and a line on the bottom. If you use any lines in the body of the table, keep them thin and light in color.

As with other elements of the poster, you need to pay attention to the amount of white space and the proximity of the data in the table. The cell margins often need to be increased (Butterick 2016). People must triangulate both across a row and down a column to make sense of

THE DATA PRISON

	Types of response			
Stimulus	On	Off	On and off	None
Positive	1,731.6	3.5	8.7	48.8
Negative	2,040.5	2.4	11.8	44.2
Both	3,552.8	0	0.9	15.9

FIGURE 9.18

Table formats.

THE ZEBRA SANCTUARY

	Types of response			
Stimulus	On	Off	On and off	None
Positive	1,731.6	3.5	8.7	48.8
Negative	2,040.5	2.4	11.8	44.2
Both	2,552.8	0	0.9	15.9

JOURNAL STYLE

	Types of response			
Stimulus	On	Off	On and off	None
Positive	1,731.6	3.5	8.7	48.8
Negative	2,040.5	2.4	11.8	44.2
Both	3,552.8	0.0	0.9	15.9

	Control group	Experimental group
Mean	20.49	32.85
Standard deviation	0.55	0.99

TABLE 9.2

Bad table, with key comparisons shown horizontally.

	Mean	Standard deviation
Control group	20.49	0.11
Experimental group	32.85	0.34

TABLE 9.3

Improved table, with key comparisons shown vertically.

any number in it. It is hard to understand the context for, and thus the meaning of, a number if the table's cells are too close together.

Choose alignment in each column according to whether it mostly contains numbers or text (Ström 2016). Columns of text should be left-aligned, like normal reading matter. Columns of numbers should be aligned along the right edge and the numbers should have the same number of decimal places, so that the width of the number quickly shows differences in magnitude. If numbers are in both rows and columns, the numbers that are to be compared should be in columns (Gelman *et al.* 2002).

TABLE 9.2 is not a well-designed table, because the means are probably the numbers that people want to compare, and the table forces them to try to compare horizontally.

TABLE 9.3 shows a better arrangement, because now the numbers being compared are in a column. In many cases, this also places related information closer together. Proximity makes comparisons easier.

Make alignment consistent in each column, including the header.

There is rarely any reason to center text or numbers in a column. An exception are headings spanning multiple columns that contain subheadings below them. Centering the heading keeps it close to more subheadings, rather than just the leftmost or rightmost. This signals that the heading applies to all subheadings.

Check what number styles your fonts have. Most software uses proportional numbers by default, in which numbers are different widths (0 is wider than 1, for instance) to make the numbers maximally compact. Many fonts also have tabular numbers that can be accessed in the software's settings. As the name implies, tabular numbers are specifically designed so that all numerals are equally wide, which optimizes comparisons between numbers in the column of a table (FIGURE 9.19). When tabular numbers are right-aligned, decimals and dividers will align vertically (Ström 2016). This only works if you use the same number of decimal places consistently throughout your table, so do that.

PROPORTIONAL LINING		PROPORTIONAL OLDSTYLE	
1,117,113.1	1,117,113.1	1,117,113.1	1,117,113.1
4,209,408.8	4,209,408.8	4,209,408.8	4,209,408.8

TABULAR LINING		TABULAR OLDSTYLE	
1,117,113.1	1,117,113.1	1,117,113.1	—
4,209,408.8	4,209,408.8	4,209,408.8	—

FIGURE 9.19

Proportional and tabular number forms.

Unfortunately, not all fonts have all these number variants. Even for fonts that do have number variants, not all software allows you to access them directly.

It's often a good idea to let the width of the data determine the size of the columns or rows, rather than spacing everything evenly (Rutter 2017).

In journals and books, the title of a table is placed above the table. This is a good practice to follow for posters, too.

Although tables are not appropriate for many posters, this is not to say that you should avoid numbers at all. As demonstrated in FIGURE 9.12, simple descriptive numbers can be shown as numbers rather than trying to make graphs of them. For example, if you want to point out that only 20% of faculty members in a department are women, putting "20% women" in large text is probably as effective as a pie chart or stack bar chart – and faster to create. A number can easily become a graphic element rather than a text element, particularly if the number is only one or two digits and set in a large point size.

Flowcharts

Flowcharts are a visual representation of processes. They are probably underused, considering how many advantages they have (Bastian 2019):

> When you create a flowchart early on, with readers in mind, you start to see things you have missed. There is something fundamentally valuable in stringently linking methods to results in this particular form. As you try to agree on the kind of sharpened, precise wording you need to communicate certain steps, you discover that you and your colleagues have been interpreting, and thus implementing, some things differently. And of course, by adding in a step that puts numbers to steps, you occasionally find a data error that may otherwise have escaped.

Flowcharts usually consists of two things: arrows and labeled boxes. While those elements are so simple that they can be made in all sort of software, there are programs dedicated to drawing flowcharts that could save you time.

Consistency is key to making a good flowchart. The arrows and boxes are to flowcharts what words are to text. Changing the style of boxes or lines in a flowchart is like changing a font in mid-sentence.

Arrows should be uniform. They should all have a single type of arrowhead, the same width, the same color, and all solid (no dashes or dots). Arrows should usually follow right-angled pathways. They need to be heavy enough to be seen, but no heavier.

Lines should not cross over on top of each other. *Exception*: you are making a deliberately confusing chart for humor. Designer Julian Hansen's flowchart-styled poster "So you need a typeface" uses overlapping lines to make the chart hard to follow on purpose to conceal the punchlines to jokes scattered throughout the poster (Hansen 2010).

Flowchart boxes come in many shapes. Some especially detailed flowcharts for computer programming have 20, 30, or 40 different box and symbol types. The most common of the specialty boxes are a rectangle to show an action, a step, or a process; a diamond to show a decision point; and an oval to show a start or stop. If you decide to use these symbols, it's probably a good idea to include a key or legend next to the flowchart.

FIGURE 9.20

Alternative flowchart designs for decision points.

USING DIAMONDS FOR DECISIONS BREAKS THE LEFT-TO-RIGHT READING FLOW!

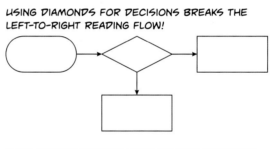

A COLORED RECTANGLE AND ARROWS IMPROVES THE FLOW!

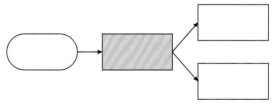

USING CONSISTENT RIGHT ANGLES MAKES DIRECTION MORE RECOGNIZABLE!

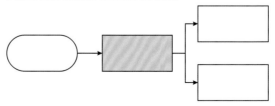

But good flowcharts can be made with rectangles and arrows alone. To use decision points as an example (FIGURE 9.20), not everyone will recognize that a diamond means a "decision." Diamonds also tend to break flow in a single direction, because they usually cause one line to veer off at a right angle compared to the other (SmartDraw Software 2019).

As with arrows, the consistency of boxes is important. The spacing between each box should be consistent. It's helpful if the boxes are the same size, too. The line widths around each box should be the same, and those lines need not be heavy ones. Boxes can be colored to help clarify certain processes, particularly if the flowchart is made mainly of rectangles (e.g., different color for processes and decisions). But try to limit the colors as much as possible. A maximum of three is a reasonable guideline (Mock 2018).

Labeled boxes can represent many different entities, which can focus attention on the overall pattern of connections. Icons, images, or photographs can be used to make the individual steps clearer. These also have the advantage of being more visual and reducing the amount of reading necessary.

In most cases, boxes should only have one arrow going to or from them. Decision points are an obvious exception, since they will have at least two arrows emerging from them. More than two arrows should be rare. If there are more than two arrows emerging from a decision point, you may need to change that single decision point into multiple decision points.

Flowcharts should flow in one direction whenever possible: either top down or left to right (FIGURE 9.21). They should not wrap around like a snake around a tree trunk. You can make an exception for cyclic processes, though.

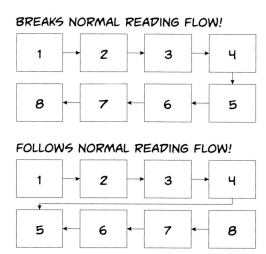

FIGURE 9.21

Flowchart reading order.

FIGURE 9.22

Clinical trial
flowchart.

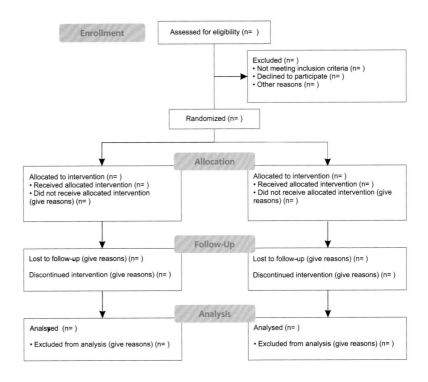

If the chart won't fit into a space, the changes in direction should follow the same conventions as normal text flow. If most boxes in a section are ordered from left to right (i.e., in a row), the next section should be placed below and the boxes continue to connect left to right. If most boxes in a section are ordered from top to bottom (i.e., in a column), the next section should be to the right.

If you have lines that cause you to return to an earlier step in the flowchart, they should be underneath the main diagram, or to the left. Long flowcharts can be broken down into multiple flowcharts.

FIGURE 9.22 is an example of a methodology chart, focused on tracking subjects through a clinical trial (CONSORT 2010).

Infographics

"Infographic" is a word that many people recognize but that is surprisingly hard to define. The word is obviously a mash-up of "information" and "graphic," but that doesn't narrow things down much. Standard graphs and charts show information graphically, but people usually don't refer to those alone as "infographics," although they may be part of an infographic. A complex graph of data might be considered an infographic, particularly if it is annotated with some simple text. Even a

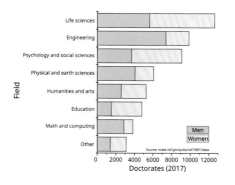

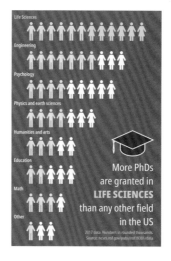

FIGURE 9.23

Bar graph of doctorates awarded.

FIGURE 9.24

Infographic of doctorates awarded.

perfectly mundane style of graph can be described as an infographic if presented in an unusual way. Take a bar chart, twist the normally straight x-axis into circle, and *voila*! Infographic.

FIGURES 9.23 and 9.24 compare data displayed in a standard bar graph and an infographic-style version. These are not necessarily the only or best way to plot these data, but illustrate the stylistic differences in approaching it: one academic and one more populist.

Because an infographic isn't a single thing, there isn't a standard way of creating one. Someone told "We need an infographic of these data" has a much more difficult task than if they are told "We need a bar graph of these data." Developing an infographic requires a clear understanding of the point you want to make. Infographics are not exploratory or democratic. They are designed to tell you a bottom line. Infographics also require you to be willing to simplify and be imprecise.

Infographics are more about their aesthetic style rather than a definable category of information display. People describe something as an "infographic" in the same way they describe a piece of art as "expressionist." It doesn't tell you anything about the content, just the look of the presentation. Since "infographic" is loosely defined, some might consider an entire poster to be a single infographic. Others might use the term for one part of a poster (FIGURE 9.25).

Given that developing an infographic is largely a stylistic exercise, it's helpful to identify some common elements of the current infographic style.

Symbols – A common part of contemporary infographic style is to use symbols to convey meaning. In theory, symbols create a sense of universality. A simple stick figure to represent a person is about as

FIGURE 9.25

Poster with
elements of
infographic style.
(Ana Maria Porras)

FIGURE 9.25

Poster with elements of infographic style. (Ana Maria Porras)

widely recognized as a symbol can get. Viennese intellectual Otto Neurath and collaborators aspired to that sort of universality. In Austria in the 1920s, they wanted to create a universally understood visual system that could help display complex social and economic facts. The result was a system called Isotype (Visocky O'Grady and Visocky O'Grady 2008), which arguably marked the start of modern infographics.

The principle that sets Isotype displays apart from graphs is that number is not directly represented by area. Instead, the repetition of symbols is used to indicate quantity, a practice that is still common in the rows of stick figures or other symbols seen in many infographics (**FIGURE 9.24**). This usually ends up approximating area, but not exactly.

There are several online archives of symbols that a poster maker can use for infographics. These include the Noun Project (thenounproject. com) for icons of many kinds of objects, Phylopic (phylopic.org) for outlines of living things, and Dimensions Guide (www.dimensions. guide) for scale drawings of a wide range of everyday objects, people, and animals.

The characteristic use of symbols pioneered by Isotype is the most contentious aspect of many infographics. But in the quest for universality, infographics often strip away too much context and become too abstract. Some objects do not lend themselves to distinctive symbols: "it seems almost impossible, for example, to design an immediately recognizable pictogram for 'potato'" (Kindel *et al.* 2009–2020). The

FIGURE 9.26

Stick figure in peril.
(Jeremy Keith)

warning sign in FIGURE 9.26 shows a person next to some sort of cliff. But the wavy lines? They might plausibly represent water, wind, or fog. Photographer Jeremy Keith was more imaginative and called this sign, "Falling into a sea of bacon" (Keith 2007).

Some symbols are deliberately abstract. The biohazard symbol of interlocking rings is distinctive, but the shape has no inherent meaning (Visocky O'Grady and Visocky O'Grady 2008). You must learn what it means, just as you learn how letters on paper represent sounds.

Some symbols go out of date. Many computer programs continue to use a "save" icon that showed a floppy disc years after floppy discs largely went away: 🖫. Likewise, phones don't look like this anymore: ☎. "Going out of date" is not the biggest problem with posters, which are designed, printed, and used within a short time frame. But the point is that symbols can be a double-edged sword.

Edward Tufte (1997) characterized the Isotype presentation of data as low resolution and not serious. The style does require rounding numbers. While this criticism has some merit, it does not change the clear utility that Isotype has for non-data displays.

Flat – Perhaps influenced by the stark symbols characterizing Isotype, modern infographics tend to be "two-dimensional." There are few graphic cues that suggest depth. That is, no photographs, no bevels, no shading, textures, crosshatching, highlights, shadows, or gradients. Objects are uniformly colored. This style has very much changed with the available technology, however. Graphs and charts drawn before the twentieth century, which would today be called infographics, made use of substantial shading, presumably because they were made from woodcuts.

Sans serif – Typography is emphasized in flat designs. Letterforms in infographics are usually sans serif (see Chapter 12). They are also often narrow and tall.

Descriptive/conclusive – Most infographics are highly annotated. Infographics do not let data speak for themselves; they tell you what to think about the data.

Chapter recap

- Many concepts can be encapsulated in one of a few basic types of visual representation.

- Simple and common graphs are often better than more nuanced but uncommon graphs.

- Tables are the worst. Seriously, just try to leave them off a poster.

- Flowcharts are an underused way of showing processes.

- Anything can be an infographic.

CHAPTER 10

Colors

Color is often the first thing we notice about an object, before details of shape or text (Dabner *et al.* 2010). We often remember the color of objects like books or CDs more than the title ("It's a book with a red cover"). Colors are important because they help break monotony. Advertisements in color are twice as memorable as those that are not (Ogilvy 1963). If you look at an academic journal, the overwhelming impression is gray. Posters are a wonderful opportunity to use color in communication in ways that journal articles rarely permit.

Filmmaker Guillermo del Toro understands the power of color. For del Toro, design is not "eye candy" – maybe sweet but probably bad for you. Instead, he calls design "eye protein." In the commentary track for the movie *Pacific Rim*, he describes how he uses color in the film to advance the narrative (del Toro 2013). Red, for example, appears only sparingly in the film. Del Toro uses red only at critical junctures in the plot, when the color is used to symbolize life. The character Mako Mori is seen in flashback as a child wearing red shoes when she narrowly escapes death. As an adult, Mori has blue hair, representing her link to the monsters (who are also blue) who almost killed her.

Use color on your poster like del Toro uses color. Deliberately. With purpose. With intent.

Using colors sounds simple but can be one of the trickiest aspects of graphic design to get right. It's easy to find a lot of claims about the psychology of color, for example that yellow rooms make babies cry. Such claims are made by people who keep their methods and results secret (Krino 1989, Werne 1989). Avoid that rabbit hole, and just have the goal of making sure that everything is visible.

Whether you create your own images or use existing ones, it's possible that some of the figures you use will heavily influence some of your later color choices. For example, if your poster is about a green tree frog, and you include photos of that frog, you're probably going to end up with a color scheme that revolves around how good things look around green (FIGURE 10.1).

FIGURE 10.1

Poster with color
palette inspired
by the frog
being studied.
(Nicholas Wu)

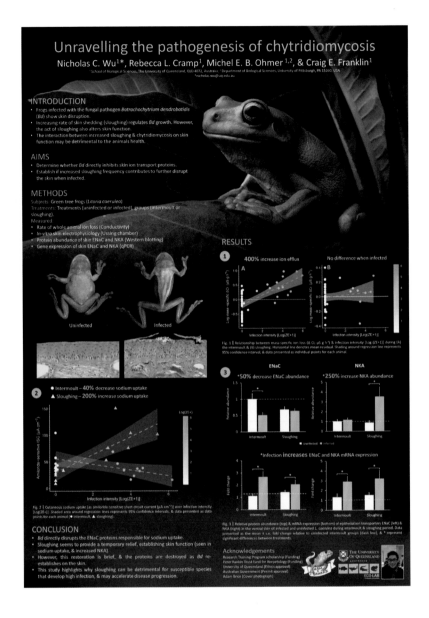

Color blindness

One advantage of leaving your poster in black, white, and gray is that color-blind people will be able to see it. If you are going to add in colors, the right choices at the start can help make your poster legible to more people.

There are several forms of color blindness, so it may be difficult to make a poster that uses colors and is equally visible to everyone. The most common form of color blindness is difficulty distinguishing red and green. Such people can usually tell blues and yellows apart readily, making these good color choices in most cases. But there is also a very rare form of color blindness which causes problems distinguishing blue and yellow.

Besides traditional color blindness, which is usually inherited and can be detected early, people lose the ability to discriminate colors as they age (Schneck *et al.* 2014).

Many computer programs have specific color palettes designed to be "color-blind safe," although they may be geared towards the more common red–green color blindness. There are color blindness simulators online that model all the different kinds of conditions. Many allow you to upload your own images. Use them to get an idea of how readable your poster is to someone who has one form or another of color blindness.

Signal differences in two ways, not just by color. If you have a color-coded graph, make the data points differ in shape, not just in color.

When making a poster, particularly graphs, it can be a good idea to start by making it entirely in grayscale. If you can interpret the graph without colors, for example, you can almost certainly make sense of the graph when you add colors in.

Color types

Color is a large and complicated part of graphic design. There is a lot of terminology for colors, but one main category is the colors of the spectrum: red, orange, yellow, green, blue, and purple. Alone, these colors are saturated, but can be made darker by mixing them with black ("shades"), or lighter by mixing them with white ("tints").

Because posters are competing for attention, new designers tend to lean heavily on saturated spectrum colors (for example, reds, blues, and greens), particularly in graphs. Red and yellow are common attention getters. There's a reason so many warning signs use red. Lime green is a lesser used but highly visible color (Visocky O'Grady and Visocky O'Grady 2008).

But the colors of the spectrum are also the colors of the rainbow. The phrase "rainbow-colored" suggests many things, but not all are positive. Yes, "rainbow-colored" suggests something bright, pretty, and attention-grabbing, but it also suggests something gaudy, garish, and simple. The more colors you use, the more likely this complaint is. Two of the most common and popular spectrum colors, red and blue, can make a particularly bad mix when placed next to together. The colors look like they are vibrating, an effect called chromostereopsis (FIGURE 10.2).

FIGURE 10.2

Adjacent red
and blue.

As mentioned above, using spectrum colors can cause problems for color-blind people.

Another large category of colors are neutral colors. These include black and white and all the shades of gray in-between, as well as off-white, creams, browns, tans, and pastels. These are safe and predictable colors. They rarely clash with each other or with the spectrum colors. They do not cause issues for people who are color-blind.

But "neutral" also implies "boring."

Good color designs on a poster are often about hitting the right balance between the attention-grabbing spectrum colors and subtle neutral colors. And you have to think about the intensity of colors, not just the type of color.

To get the right balance, let's revisit the concepts of visual hierarchy and visual weight. Color is part of the visual hierarchy discussed in Chapter 7, and just as you don't want to use bold and underline to emphasize a piece of text, you don't want to use color that results in overkill. As we saw in Chapter 7, there are many interacting cues that determine how much visual weight an object has.

If you are using color spread out over a large area, such as to define a column on the poster, the color can be very light and pale, but it will still visibly be that color. It will still "read" to the eye.

Bright colors from the spectrum have a lot of optical weight. A small, brightly colored point on a poster is like a neutron star: it's tiny, but it pulls you towards it because it has so much visual weight. As the color becomes lighter and less saturated in hue, it loses visual weight (Landa *et al.* 2007).

Which way would you expect the seesaw in **FIGURE 10.3** to tip?

Small spots of color, like data points on a larger graph, can be very intense and saturated to stand out and draw attention to themselves. Very large

FIGURE 10.3

Demonstration of
visual weight.

CONTRAST EMPHASIZES WHAT IS IMPORTANT

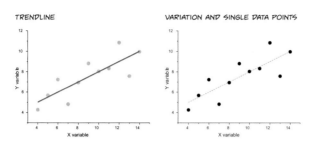

FIGURE 10.4

Scatterplots that either emphasize trendline (left) or variation (right).

objects should be in very muted colors to prevent them from dragging attention away from the rest of the poster.

You can use visual weight in graphs to keep the focus clearly on a summary while still including a lot of data. For example, using a bright color and slightly thicker line will bring most of the focus to the regression line (FIGURE 10.4). But plotting individual data points in light gray makes them still visible, particularly up close if someone is making a detailed inspection.

Another consideration that makes color use tricky is that we interpret the same light stimulus very differently depending on the context (FIGURE 10.5). There are tons of color illusions that show how we can see the same color value on a computer screen very differently depending on what is next to it. Whether we think an object is in a well-lit or darkly lit area completely changes our interpretation of colors. A well-known example was the viral photograph simply called "the dress," which showed a dress that looked white and gold to some people, but blue and black to others. A light object can look even lighter next to a darker object but will look darker when placed next to a light object. Again, design isn't math.

Designers often take advantage of how the context surrounding a color influences the perception of it by working with complementary colors. These are colors that "pop" when placed next to each other (see FIGURES 10.8 and 10.9).

THE SAME ORANGE THAT FADES NEXT TO RED WILL POP OUT NEXT TO BLUE!

FIGURE 10.5

Perception of colors changes with context.

Color palettes

One of the best ways to get your poster looking great, particularly from a distance, is to choose a color palette: a limited set of colors that are used on most, if not all, of the elements on your poster. Here is a suggested process (Stone *et al.* 2006).

1. Choose a main color. One logical way to pick a main color is to draw from a photograph of something related to your project. As mentioned above, if your research is about a green tree frog, "green" is a sensible main color (FIGURE 10.1). Many people base poster colors on their institutional colors (e.g., their university's logo). Or you can pick a main color from some object, picture, or artwork that has colors you like, since you already know those work together. If you like skateboards, sample colors from a skateboard. You can also pick main colors from cultural connotations. For example, if your poster features China, you might not want to use greens and blues, since the association with the country's red flag is so strong. If your poster is about food, you may not want it to be blue, because there are few blue foods. (Look at logos for fast food chains and see how many prominently feature blue. There are very few compared to red and yellow logos.)

2. Decide how many main colors you want. You may decide to have your poster use variations of your main color (monochromatic design). Two or three colors is enough. Balancing four or more colors is challenging even for experienced designers, and you start to lose the benefits of having a palette.

3. Choose accent colors (if any). Accents are secondary colors that will take up less space in your design and will probably be used for emphasis or highlights. Remember that a color that contrasts with most of the others can have a lot of visual weight, even if quite small. Again, if you are basing your color scheme off something that is already existing (a photo, a logo, etc.), this decision might be already made for you. If not, picking color harmonies from a color wheel (below) is a simple and effective way of choosing accents.

4. Create shades of the main and accent colors. You can achieve a lot of variation simply by making colors lighter or darker. Often, using a very light tint of your main color instead of white or a very dark shade of that main color instead of black creates a more coherent appearance.

5. Test the colors. Some combinations that work in theory may need significant work to play nicely together in practice. The only way to tell is to put them on the page and see.

COLOR WHEEL FOR INKS!
(ALSO KNOWN AS THE SUBTRACTIVE ARTISTS'
PRIMARY COLOR WHEEL, OR RYB COLOR WHEEL)

FIGURE 10.6

Color wheel for inks.

When picking colors for a palette, it is incredibly helpful to use a color wheel. There are many ways to arrange color, but the example in FIGURE 10.6 uses twelve colors of ink (subtractive colors) for ease (Stone *et al.* 2006).

The primary and secondary colors sit at an equal distance from one another (FIGURE 10.7). Secondary colors are equally distant between two primary ones.

If you are going for a palette of two colors, try colors that oppose each other on the wheel, which are complementary colors (FIGURE 10.8). These colors have a lot of contrast, making for bold designs. Blue and orange is a favorite (FIGURE 10.9). Its use in movie posters is practically cliché. Purple and yellow are more difficult to work with, perhaps because there

PRIMARY COLORS   SECONDARY COLORS

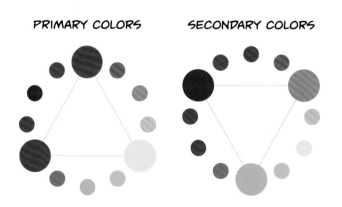

FIGURE 10.7

Primary and secondary colors positioned on color wheel.

FIGURE 10.8

Complementary
colors.

FIGURE 10.9

A poster using
blue and orange
as contrast colors.

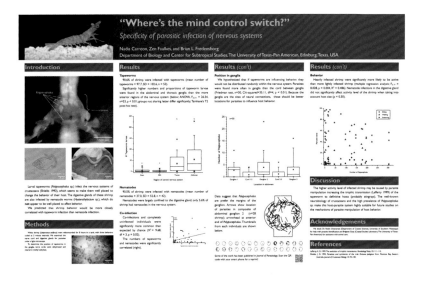

are not many things in the natural world that are purple. Red and green
can be hard to use in many countries because that color combination is
so strongly associated with Christmas. But it is also worth noting that a
red rose against green leaves is an example of this pair of complementary
colors appearing in nature.

If you expand your palette to three colors, there are a couple of options.
A triadic color scheme (FIGURE 10.10) picks three colors equally distant
on the wheel. This can run the risk of looking loud and confusing.

An analogous color scheme (**FIGURE 10.11**) uses three colors adjacent to each other on the wheel. The effect can be close to monochrome.

Split complementary colors are a compromise of sorts (**FIGURE 10.12**). Three colors are used, with two adjacent to the complement of the third. In this scheme, the two colors closest to each other can serve as the main colors on the poster, while the third provides contrast.

Four colors are difficult to work with, but one possible scheme is double complementary colors (**FIGURE 10.13**).

Understanding these color harmonies is a useful way to think about options. You can ask whether you want a poster with contrasting colors (complements or split complements), or something with more cohesion

TRIADIC COLORS ARE EVENLY SPACED AROUND THE COLOR WHEEL. THESE PROVIDE CONTRAST!

ANALOGOUS COLORS SIT SIDE-BY-SIDE ON THE COLOR WHEEL. THESE PROVIDE UNITY!

FIGURE 10.10

Triadic colors.

FIGURE 10.11

Analogous colors.

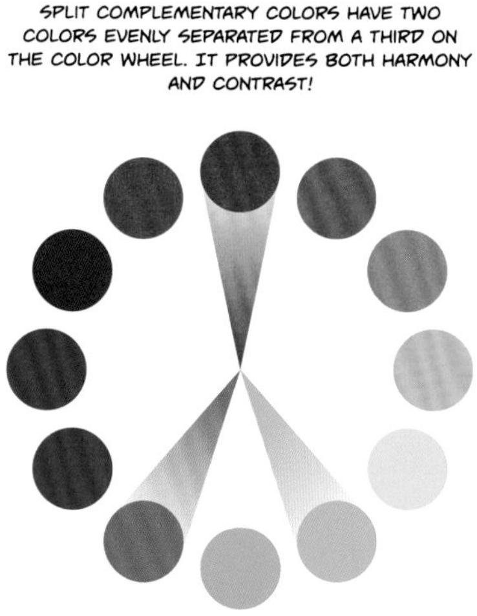

SPLIT COMPLEMENTARY COLORS HAVE TWO COLORS EVENLY SEPARATED FROM A THIRD ON THE COLOR WHEEL. IT PROVIDES BOTH HARMONY AND CONTRAST!

DOUBLE COMPLEMENTARY COLORS PROVIDE MORE SIMILAR TONES, BUT AT THE COST OF MORE COMPLEXITY!

FIGURE 10.12

Split complementary colors.

FIGURE 10.13

Double complementary colors.

(monochrome, analogous, split complement). Many websites can perform color selections using these principles for millions of colors, not just the twelve shown here. Search for "color palette generator" for these tools.

A major advantage of using computer-based color palettes is the ability to easily explore new color combinations. The examples here focus on the spectrum colors, but many color palettes move away from simple saturated colors to complex shaded colors. You may need to think not just "green" but whether that green is "moss," "sage," or "seafoam." Or what colors might best approximate metallics like gold or silver.

You can usually add grayscales to any color palette without too much disruption. Neutral colors are also less prone to great conflicts because of their lower intensity. Grays can be the glue that holds a design together. As noted above, plotting most data in neutral colors like gray makes it possible to highlight important points using color.

Color consistency

The joy of colors is that they are so variable and rich. But the frustration of working with colors is that they are so variable and rich. When designing, saying "I'd like this poster to be blue" leaves a lot of room for maneuver. There is a vast difference between navy blue and sky blue, and sometimes design requires matching colors exactly.

Most computer graphics programs have an "eyedropper" tool that allows you to select a color from an existing image. Pulling colors from your pictures using the eyedropper tool is one of best ways to unify the colors on your poster.

For example, most posters designed on a computer start with pure black text on a pure white background. But when you look at a physical book or magazine, the paper is probably not perfectly white, and the ink is not perfectly black. You could sample the darkest part of your picture and use that in place of black, and the lightest part of your picture could be used in place of white. The brightest colors might be suitable for use as highlight colors in graphs or charts.

Because digital photographs provide so many colors, use the eyedropper tool rather than eyeballing color matches. For example, dark areas on photographs are rarely straight black. It's more likely that the darkest parts of your photograph are very dark brown, dark blue, or something else. But software tends to show simple colors, like pure black, that you don't always find in photographs. Because black is neutral, a pure black might look fine, but using that same very dark brown that is in your picture will tie the poster together in a way that the pure black won't.

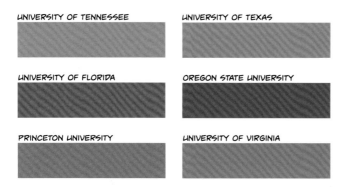

UNIVERSITY OF TENNESSEE

UNIVERSITY OF TEXAS

UNIVERSITY OF FLORIDA

OREGON STATE UNIVERSITY

PRINCETON UNIVERSITY

UNIVERSITY OF VIRGINIA

FIGURE 10.14

Shades of orange
used by universities.

Institutions are notoriously fussy about their colors. FIGURE 10.14 is a
collection of specific shades of orange used by American universities.
Institutions even nickname their oranges ("Rotunda Orange," "Beaver
Orange").

The oranges in FIGURE 10.14 were all found in institutional style guides
that specified the exact colors for use. Some can (and should) specify
colors in different ways, because different people have different purposes.

Pantone is a company that has made a business out of color. They have
an extensive set of tools for matching colors and create products specifically in desired colors. Many style guides include Pantone specifications,
and many art products can be purchased in Pantone matched colors.
For example, if you need a marker and colored paper that both match
your institution's shade of purple, you can often do this by finding and
matching the appropriate Pantone code.

Screens versus ink

A final complication is that colors work differently on a computer screen
than on paper. There is a huge difference between seeing light projected
through pixels on a screen and seeing light reflected by ink on a page. The
visibility of ink on paper depends on the available ambient light in the
poster session room, and it's not always good. Poster sessions are often
held in rooms with dim artificial light that is not under the control of the
presenter, so you must design your poster with a safety factor so it can
stay visible even under poor lighting conditions.

Computer screens generate light. White things will look bigger because
they have a glowing halo emanating from them. On a computer screen,
the primary colors that can be used to make any other colors are red,
green, and blue. When you combine all three of these colors, you get
white, so it is called an **additive** mix. This color space is abbreviated RGB

in most color settings. Notably, PowerPoint, the most common software used for posters, works *only* in RGB.

Poster paper reflects light. Black things will look bigger because they are printed in ink that bleeds outward. On paper, the primary colors that can be used to make other colors are cyan, magenta, and yellow. When you combine all three of these colors, you get black, so it is called a **subtractive** mix. Rather than always making black by using these colors, printers usually add a separate ink for black (abbreviated K, which coincidentally avoids confusion with blue), so this color space is abbreviated CMYK. Surprisingly, using 100% black ink may look like the darkest possible black on a screen but can look more like dark gray when printed (Vlahos 2018). Darker, richer blacks are made by mixing in the other three colors.

If you have a printer in mind, check to find out if they have any preferences for color space. Some printers prefer to receive original files in CMYK to ensure greater fidelity between what you see and what you get. If your printer wants CMYK files, your software may allow you choose whether to work in RGB or CMYK. Chapter 22 covers more on working with printers.

Chapter recap

- Color is an accessibility issue. Pick colors so that your poster can be understood by someone who is color-blind.

- Color is powerful but easy to misuse.

- Color theory can help you pick a limited color palette to bring cohesion to the poster.

- You can ensure different components on the poster are consistently colored using eyedropper tools.

- Colors on screen are made of light and can look different from colors on the page made with ink.

CHAPTER 11

Beyond paper

Poster sessions are usually *laissez faire* events. Because there are no poster police, you can do a lot besides just putting up a piece of paper. You can create displays that are multimedia and more interactive. But these do require more planning up front.

QR codes

A quick response (QR) code is a square computer graphic that encodes some simple digital information, usually a link to a website. It can also encode other sorts of information, like plain text or an image, but links to websites are the most common. An audience member can take a picture of a code on a smartphone, which will often recognize the code and open the link – often without having to download a specific app. A QR code gives your audience a fast way of accessing a website without having to type in a long URL.

The kind of information you might want to encode using a QR code is anything that is useful but not necessary to understanding the poster. A QR code could link to online videos, a data set, a preprint, or a complete author list (if the list is exceptionally long).

The QR code format has a lot of redundancy built into it (FIGURE 11.1). You can take pictures of the code from a wide range of angles and distances.

THE APPEARANCE OF QR CODES CAN BE CHANGED WITHOUT AFFECTING HOW THEY WORK. THESE ALL LEAD TO THE BETTER POSTERS BLOG!

FIGURE 11.1

QR codes.

FIGURE 11.2

Signposting
QR codes.

SAY **WHY** PEOPLE SHOULD SCAN THE CODE AND
PROVIDE **OPTIONS** FOR GETTING THAT CONTENT!

 Scan to download the full text of
Stinging the Predators!

Also available at: http://bit.ly/StingPred

The code can even be changed or obscured, and it will still work. This presents some design opportunities, because you can make a QR code blend in with the colors and shapes of your poster.

In theory, using a QR code to direct viewers to a website is quick and simple. But in practice, using a QR code is a barrier to getting to a website. It requires that someone recognize a QR code and how it works, have a phone with them, have that phone with a decent amount of charge in its battery, fish it out of their pocket or purse, and take a picture. These are all steps that someone just reading the poster in a busy conference hallway does not have to take. Remember that every bit of effort you ask people to put in decreases "relevance" to them. And on top of that, some people just don't like QR codes on principle.

If you put up a QR code, give people a reason to scan it. Make sure you indicate what people will get when they go there (FIGURE 11.2). For example, "Take a picture to download the full paper."

The distinctive look of a QR code can make people think, "Oh, there's more to see online" even if they don't plan to scan the code. Sometimes people will prefer to type in a short web address rather than take a picture of a QR code. When posting a QR code, provide a short URL too (search for "URL shortener" on the web). Another advantage of a plain URL is that people may not be able to use the code right there at the poster because they have no internet connection. Many conferences are in basements or locations where there is no wifi or phone reception is poor. A note or picture of the URL makes the link available later.

Video and moving images

Sometimes you want to show information that is dynamic, not static. The traditional way of showing movement on a page is to grab a series of individual frames and present them in sequence. But there are other ways to bring moving images to a poster.

Tablets and phones

If you place a video online, you can use QR codes or short URLs to link out to your video.

But rather than relying on other people's phones, you can show videos to people on your own portable electronic device. Tablets work better for showing videos in a poster session, because the screen is a little bigger. Download the video before your session, in case the poster session is in a room with no connectivity.

Instead of holding a tablet in your hand, you might plan on mounting it on the poster. Dedicate a space for the tablet in your layout. Measure your tablet, then block out a plain white space where it will sit on the poster. Tablets are light enough that they can be held up by tacks if the poster board is in good shape. You can also use hooks or strings to hold the tablet on the board. There are also commercially available wall mounts for some of the more popular tablets.

Flipbooks

An alternative to video displays is to make a flipbook. This low-tech solution has an almost irresistible charm. People love to flip pages! A tactile booklet invites people to come up and interact in a way that scanning a QR code or pushing a button on a tablet does not.

Some businesses will take a video clip and turn it into a printed flipbook. But if you are patient, you can do this yourself by grabbing frames from the video, printing them, and binding them together with some glue.

Interactive displays

Because poster sessions give you a little space and a lot of freedom, there are plenty of things you can add to your poster that invite the audience to do more than just read it. You can add objects that entice people to physically come and touch the poster.

Commentary

You can change your audience members from passive viewers into active participants by creating space for comments.

Spenser Babb-Biernacki created a billboard-style poster (FIGURE 11.3) with about a third of the poster left blank, except for a question at the top (Babb-Biernacki 2019a, 2019b). The poster was laminated with a special coating used for dry erase markers. A couple of markers underneath the

FIGURE 11.3

Poster with space
for comments.
Top: start of session.
Bottom: end of
session. (Spenser
Babb-Biernacki)

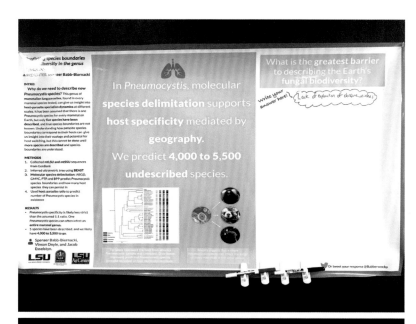

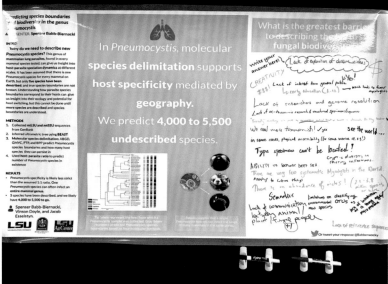

poster made it possible for people to leave their answers to questions. When the space was full, he took a picture of the comments, erased the question and answer space, and started the process over again.

Another way to do this that does not require special lamination is to leave space on the poster for sticky notes (FIGURE 11.4). Label the space with something like, "Leave a comment here!" Take a small container that can be attached to the poster board, some pens, and sticky notes. Attach them next to the comment space on the poster. Because people are often

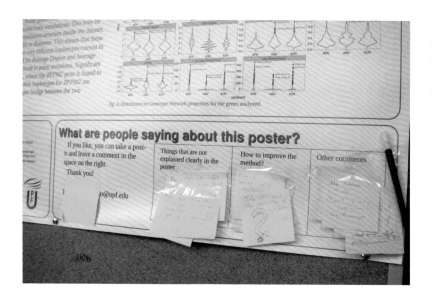

FIGURE 11.4

Inviting feedback
using sticky notes.
(Giovanni Dall'Olio)

intimidated by empty spaces, ask one or two friends to contribute the first couple of notes.

To get people to participate, you must convey that you take what they say seriously. It's important to make these sorts of spaces look intentional. A single piece of letter paper tacked up next to a pencil on a string looks sloppy, unplanned, and uninviting.

Papercraft

Some posters use papercraft, like the kind used in pop-up books.

For instance, you can have covers that people lift to see what's underneath. This invites interaction, making it more like a museum exhibit than a page to be read. There are a couple of ways to achieve this. You can print a question on one piece of paper and have the answer on the main body of the poster. A little transparent tape at the top of the "question" page, and you have a flap that can be lifted.

But there is a more elegant solution that leaves no sticky tape exposed. Monica Granados designing her poster with spaces that said, "Lift for figure." She printed her poster on heavy paper, then used a precision blade to make three cuts on the left, right, and bottom of the panel, leaving the top to act as a hinge. She then printed the figures she wanted to reveal on a smaller desktop printer, and glued them underneath the flap (Granados 2018).

You can generate movement by creating tabs that people pull, or by pinning a piece of heavy paper onto the poster with a tack so that the paper can be rotated.

Three-dimensional images

Some data greatly benefit from being shown in three dimensions, like some complex structures recorded with confocal microscopes or computerized tomography (CT) scans. There are several ways to show three-dimensional images.

Side-by-side stereograms (FIGURE 11.5) show two slightly different images side by side. Some people are reasonably skilled at crossing their eyes so the two images overlap, but there are also glasses that can help people see the 3-D effect. An advantage here is that the slightly different views are understandable even if you can't see the image in 3-D. In software that creates 3-D graphs, you can get a reasonable approximation by creating an image of one graph, exporting it, rotating it about 6 degrees, and capturing the second image. Two photographs taken from different positions in the same horizontal plane can be used to create 3-D images.

An anaglyph (FIGURE 11.6) is a single picture that superimposes two views, one view colored to enhance red and the other view colored to include more blue or green. Here, you need special glasses to get the 3-D effect. Just make sure you have enough of the 3-D glasses for several people to see the images at the same time, without having to wait. The effect works best if the poster is printed on matte paper, because glare interferes with the effect.

The newest way to show 3-D data is to use 3-D printing. Some people scan small items, enlarge them, and print them larger so they can be seen more easily. There are smartphone apps to scan small items in 3-D. These tangible objects might explain complex structures better than visuals. Depending on the objects, you might be able to attach them to the poster or have them on hand to show to audience members. Solid objects can also make your poster content more accessible to people with visual disabilities.

FIGURE 11.5

Stereogram created by rotating graphs.

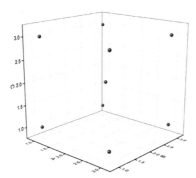 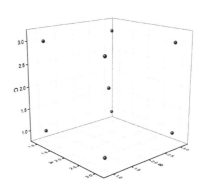

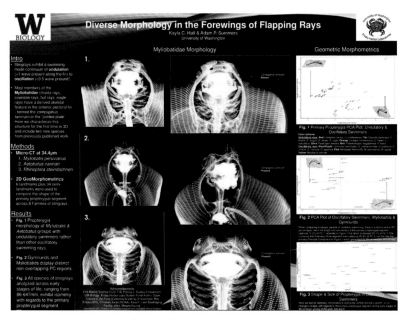

FIGURE 11.6

Top: poster using anaglyphs to create 3-D images. Bottom: poster creator Kayla Hall presenting the poster. (Kayla Hall)

Texture

Any poster will have a texture as well as a look. Shiny, glossy paper is smooth but hard. Fabric looks softer and rumpled. With a little adhesive or tacks, you could add different kinds of things to a poster that people could touch.

Admittedly, there are probably not many cases where you need people to touch things to make a point. But think of how cool it would be if you did.

FIGURE 11.7 shows a theatrical movie poster that could have just been paper but … it's furry! How can anyone look at this and not want to walk up and touch this thing? If it had just been paper, it would be just another poster. Add an unexpected element, like the sense of touch, and it becomes something worth talking about.

FIGURE 11.7

Poster for the movie *Abominable*.

Painting

For the most part, this book assumes that you are going to design a poster on a computer, and print it using a large-format printer. But one person turned a poster into an artistic canvas in the most literal sense. Karmella Haynes painted her research project in acrylics and presented it as a poster at an international meeting on synthetic biology (Eisen 2011a, 2011b).

Chapter recap

- Poster boards are not limited to displaying paper.
- Online resources, video, and using the board for other displays provide opportunities to invite interaction with the audience.

CHAPTER 12

Text and type

Academics tends to care a lot about what words *say* but tend to forget how words *look*. Even if you have some awareness of type, you may be likely to overlook the impression that words have when laid out on a page.

Typography terminology

Typography is a specialized craft with its own unique terminology (Lupton 2004), which is often misused or misunderstood. The most common point of confusion, wrought by the wording used in computer software, is the meaning of "font" (**FIGURE 12.1**) A typeface is a design of letter forms. A single typeface can have many fonts. The typeface is the design, the font is the package it comes it.

The difference between a typeface and a font is like the difference between a song (a piece of design) and a performance of a song. "Light My Fire" is the same song whether performed by The Doors or José Feliciano. A vinyl record, compact disc, cassette tape, or digital download (a way to get the thing) are almost the musical equivalents of fonts. The typeface is usually the more interesting thing; the font is just the container it happens to be in.

The most common file format for fonts are OpenType and TrueType, which are supported by both Windows and Macintosh computers. OpenType is the more recent format, and supports many features that TrueType does not.

FIGURE 12.1

The difference between typefaces and fonts.

DIFFERENT TYPEFACES	DIFFERENT FONTS
Caecilia	**Noto Sans Black**
Conqueror Sans	Noto Sans Regular
Sugo	Noto Sans Condensed
Cherry Blossoms	Noto Sans Condensed Thin

SAME TEXT, 12 POINT, TWO DIFFERENT FONTS

Lorem ipsum dolor sit amet, consectetur adipiscing elit. Nullam nec sapien a ex convallis scelerisque nec a mi. Pellentesque lacus massa, porta vitae sapien quis, placerat vehicula tellus. Maecenas est mauris, vulputate et fermentum quis, pellentesque aliquet velit. Etiam vitae lorem a nunc suscipit tempor. Nulla eleifend dapibus ipsum vel tristique. Suspendisse sit amet dolor non nunc interdum elementum. Curabitur eget aliquet felis, a sollicitudin ligula. Sed nisi ex, eleifend ut urna ac, ultrices elementum mauris. In rhoncus finibus dolor, eu vulputate lacus hendrerit in. Fusce non pulvinar lacus.

WE'RE DONE HERE, BUT STILL GOING STRONG OVER ON THE RIGHT!

Lorem ipsum dolor sit amet, consectetur adipiscing elit. Nullam nec sapien a ex convallis scelerisque nec a mi. Pellentesque lacus massa, porta vitae sapien quis, placerat vehicula tellus. Maecenas est mauris, vulputate et fermentum quis, pellentesque aliquet velit. Etiam vitae lorem a nunc suscipit tempor. Nulla eleifend dapibus ipsum vel tristique. Suspendisse sit amet dolor non nunc interdum elementum. Curabitur eget aliquet felis, a sollicitudin ligula. Sed nisi ex, eleifend ut urna ac, ultrices elementum mauris. In rhoncus finibus dolor, eu vulputate lacus hendrerit in. Fusce non pulvinar lacus.

FIGURE 12.2

Column space used by different typefaces with the same point size.

Type (and line thickness) is measured in points. There are 72 points to an inch – in theory. In practice, point size gives only a rough guide as to how much space text will take up (**FIGURE 12.2**).

Very rarely, you may see references to "picas." A pica is 12 points, so there are six picas in an inch.

There is also special terminology to describe the proportions of letters by comparing the lowercase letters to their extremes (**FIGURE 12.3**). A typeface with a large x-height, or long ascenders and descenders, needs more space between each line. Large x-height makes blocks of text "gray." Long ascenders and descenders risk touching each other (Dabner *et al.* 2010).

FIGURE 12.3

Terms used to describe typefaces.

Two of the most common categories for typefaces are "serif" and "sans serif" (FIGURE 12.4) Serifs are little flares arising from the shapes of letters, usually at the ends of lines. *Sans* means "without" in French, so sans serif typefaces lack serifs.

The typical claim is that serifs make letters more distinctive and easier to read when set in long sequences of small text (like books), but that sans serif makes letters easier to read when set in short sequences of large text (like posters). In fact, many people and books and websites that tell you with great certainty that you should use sans serif typefaces for posters (Briscoe 1996, Duarte 2009a, Alley 2018). These all claim that sans serif is more legible for tasks where the words are few, the print is large, and distance to the reader is great. And that pretty much describes posters to a T. For example, "A serif type can be distracting, especially in a large title" (Briscoe 1996), but why this should be so is, sadly, never explored or explained. Even those that advocate serif typefaces for the main text will suggest using sans serif type for big text (Hess *et al.* 2013, Purrington 2019). Surely all those smart people must be on to something.

But posters don't always need to use a sans serif typeface.

> Typography is not a science. Typography is an art. There are those who'd like to "scientificize"; those who believe that a large enough sample of data will somehow elicit good typography … A lot of time is wasted attempting to prove that [sans serif or serif] is better than the other for setting extended text. (Boardley 2008)

FIGURE 12.4

Serif and sans serif typefaces.

SERIF TYPEFACES

Times New Roman
Sphinx of black quartz, hear my vow!

Bodoni
Sphinx of black quartz, hear my vow!

Courier Prime
Sphinx of black quartz, hear my vow!

Sitka
Sphinx of black quartz, hear my vow!

Noto Serif
Sphinx of black quartz, hear my vow!

Calluna
Sphinx of black quartz, hear my vow!

SANS SERIF TYPEFACES

Arial
Sphinx of black quartz, hear my vow!

Verdana
Sphinx of black quartz, hear my vow!

Futura
Sphinx of black quartz, hear my vow!

Cabin
Sphinx of black quartz, hear my vow!

Noto Sans
Sphinx of black quartz, hear my vow!

Calluna Sans
Sphinx of black quartz, hear my vow!

People cheerleading efforts to make typography scientific will sometimes say things like, "These recommendations are not opinions, but rather facts based on research done by academics following rigorous protocols" (Kammeyer 2009). This sounds impressive until you look at the research, and realize it often involves showing a small number of typefaces to a small number of participants (Tufte 2002, Mackiewicz 2007). Given that there are thousands of typefaces that have been designed over the centuries, a paper that tests ten is not particularly convincing stuff to support a conclusion of Typefaces Thou Shalt Use™.

For posters, the legibility of text is mainly determined by the size of the text. If you make it big enough, people will be able to read it.

Modern fonts can have many features that are not obvious to beginners. These include swash capitals, tabular numbers, stylistic alternatives for certain letters, and sometimes a variant of the entire alphabet (FIGURE 12.5).

These features are not always easy to find. The OpenType font specification supports them, but not all type designers bother to create them. Different software packages have different levels of support for them. This complicates matters when working with multiple software programs (which is common when making posters), because a carefully chosen alternative visible in one application may disappear when imported into another application (see *Checking for errors* in Chapter 22).

STYLE SET 1 brilliant

STYLE SET 2 brilliant

DEFAULT NUMBERS 1926

OLDSTYLE NUMBERS 1926

DEFAULT CAPITAL *Superb*

SWASH CAPITAL *Superb*

DEFAULT CHARACTER *wonder*

FIRST ALTERNATE *wonder*

SECOND ALTERNATE *wonder*

FIGURE 12.5

Examples of swashes and alternative letterforms.

For the most part, you will probably rarely use these extended features. But it's helpful to know that they exist, so that you can look for these options, and can use them if you want to. Little details like using swashes or tabular letters are touches that can elevate a poster above the ordinary.

Dingbats are icons or shapes that can be used in typesetting. These are what typesetters used before emoji.

Typeface selection

Typefaces make a huge difference to poster's readability and overall impression. Using type well is one of the most versatile and useful skills in a poster designer's skill set. FIGURE 12.6 gives one example of the power of typography. The typeface alone completely changes the tone and perception of the message. People strongly agree on the personalities expressed by different typefaces (Brumberger 2003).

Because almost all posters are designed on desktop computers, one of the easiest holes to fall into is to use the standard fonts that were shipped out with the operating-system software on your computer. There are enough fonts on most desktops to give the illusion that you have a lot of choices, but software packages usually default to some very common font. And once a font is preselected, people rarely change it. It's easy to fall into a "default" mode and use the same fonts over and over without really making any explicit decisions of your own.

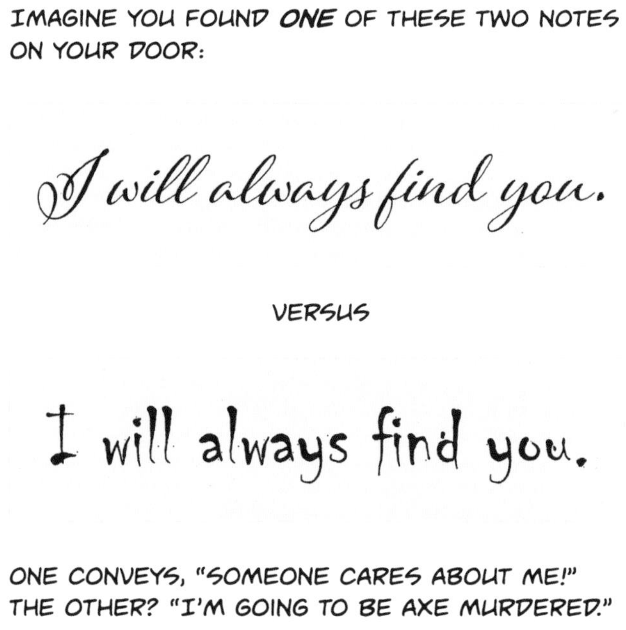

FIGURE 12.6

Example of emotional cues in typefaces.

IMAGINE YOU FOUND **ONE** OF THESE TWO NOTES ON YOUR DOOR:

I will always find you.

VERSUS

I will always find you.

ONE CONVEYS, "SOMEONE CARES ABOUT ME!"
THE OTHER? "I'M GOING TO BE AXE MURDERED."

Williams (2004) suggests that when you're starting to design, do not use a system default or common typefaces like Times New Roman, Helvetica, Arial, or Calibri for a while. It's not that those typefaces are bad. They are very good typefaces for many tasks, which is why they are software defaults. But they have been used so much for so long that even when used appropriately, they can look bland or dated. Helvetica's reputation is undiminished among graphic connoisseurs, but it has been used so often that it nearly became synonymous with standardized, depersonalized corporate imagery (Hustwit 2007). Seth Godin (2010) wrote, "If you send me a flyer with dated, cheesy or overused type, it's like showing up in a leisure suit for a first date."

Type styles change so much that you can often pinpoint the decade a film, television series, or advertisement debuted by the kind of type that's used, and many typefaces have been around a long time. Times New Roman was designed in the 1930s, Helvetica in the 1950s, and Arial in the 1980s. Make sure your poster looks like it's from this century – unless you've explicitly decided you want a retro look!

Once you've put aside the defaults as options, you can start looking for a new typeface that meets your needs.

We already talked about some of the needs. Posters need to be easy to read, visible at a glance from a distance while the viewer is moving, potentially in bad viewing conditions such as dim lighting. They also need to be accessible to people with visual problems, whether color blindness or poor ocular focus.

One of the best ways to make a good design is to look at how similar problems are solved by common, tried-and-tested objects. The set of problems just described are also faced by highway signs. Highway signs use large, high-contrast, sans serif typefaces with uniform line widths. The type is rarely set in condensed or expanded fonts, which are less familiar and increase the likelihood of misinterpretation. You see similar sorts of type used in places like airports, public transportation terminals, and so on. In the United States, the Americans with Disabilities Act (ADA) also frequently requires the use of sans serif typefaces for accessibility purposes (Department of Justice 2010).

Academic work often has complex typographic needs. You may need superscripts, subscripts, Greek letters, mathematical symbols, various accents and umlauts and diacritics, and more. You need to check that the font you want to use has the symbols you need to typeset your poster.

If you use something other than the fonts that came with your computer, it will give the poster a distinctiveness that can serve well. There are thousands of fonts out there to choose from. If you're looking for a way to set your poster apart from the pack, splurge on some fonts. Buying some fonts that 95% of the other presenters won't have on their computers is a good investment, and is one of those little subtle tweaks that helps to move a poster from good to gorgeous. A few of the many font shops online include Fontshop (www.fontshop.com), MyFonts (www.myfonts.com), K-type (www.k-type.com), and Hoefler & Co. (www.typography.com).

Because you should have an idea of the subject of your poster, one of the most helpful things you can do in trying to pick a typeface is to try a lot of them. And don't just look at some sample. Look at the words that will appear most often on your poster. See how your title and common words look in that font.

WALRUS octopus Wall·e The Taking of Pelham 123
– Calluna Sans

WALRUS octopus Wall·e The Taking of Pelham 123
– Century Gothic

WALRUS octopus Wall·e The Taking of Pelham 123
– Noto Serif

WALRUS octopus Wall·e The Taking of Pelham 123
– Garamond

For instance, let's say your poster is about walruses. In the samples above, notice the gaps between letters in "WA." That's a pair of letters that often benefits from kerning. It's an issue in all of them but seems worse in sample #2. It can be addressed (discussed in Chapter 20), but you have to be aware it needs addressing.

Maybe you have a gene called *quill 1I*.

quill 1I   – Garamond

quill 1I   – Gill Sans

quill 1I   – Franklin Gothic

quill 1I   – JT Marnie Light

quill 1I – Univers

quill 1I – Futura Bk

quill 1I   – Georgia

Notice how the different typefaces do better jobs of distinguishing the lowercase L, the number 1, and the uppercase I. Some clearly distinguish all three (e.g., Georgia), but these three are indistinguishable in others (e.g., Gill Sans). That may not matter for your text, but it could be a big problem for some other text.

"What about Comic Sans?"

Ah.

Academics have an inexplicable fondness for Comic Sans. In 2012, researchers at CERN announced they had discovered the Higgs boson (Gianotti 2012). This was a long-predicted particle. The hunt for this particle was the primary reason for building the Large Hadron Collider, the largest and most complex machine humans had ever built. The discovery of the Higgs boson was arguably the most important advance in physics in maybe fifty years. But they made that announcement on slides with Comic Sans (FIGURE 12.7).

Some of the world's top scientists were widely mocked for this. And rightfully so. Among designers and comics fans, Comic Sans isn't so much a typeface choice as the punchline to a joke.

Some academics think that people who dislike Comic Sans are fuddy-duddies with sticks up their rear who are against any hint of playfulness. (There is some truth to this.) What is wrong with a little fun?

Nothing is wrong with fun.

The problem with Comic Sans is not that it is a comic style typeface. You've no doubt noticed that there is a comic typeface used extensively throughout this book. The problem is that Comic Sans is a bad example of the style. Comics retailer Lee Hester knew that everyone recognized this (Klein 2009):

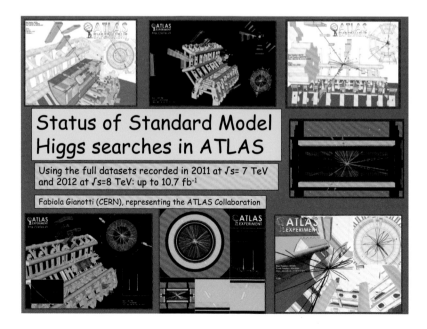

FIGURE 12.7

Slide announcing the discovery of the Higgs boson. (Fabiola Gianotti)

"How can it not be for comics, if it's called COMIC SANS?" they cry.

"Have you ever seen a comic book, or have you ever looked at a comic strip like the ones they run in YOUR OWN NEWSPAPER?" I say. "Does any of the lettering you see in those strips remotely resemble Microsoft Comic Sans?" I ask.

"Well, er, no," they allow.

This is apparent as soon as you see classic comic-book phrases in a professionally created typeface and in Comic Sans (FIGURE 12.8). Comic Sans leeches away the power of the classic comic phrases.

There are long-standing conventions in comics that Comic Sans doesn't follow. Comics traditionally use only uppercase letters. (This tradition has been weakening since most lettering is done digitally rather than by hand.) Comic Sans has lowercase.

But even the uppercase letters in Comic Sans have problems. The uppercase letter I in comics comes in two forms. There is an I with crossbars. That's reserved for the first-person personal pronoun. Every other I is shown without crossbars. But in Comic Sans, every uppercase I has crossbars. "It looks completely wrong to the comic eye," said artist and letterer Dave Gibbons, whose work was badly imitated in creating the typeface (Schofield 2009). Gibbons created a self-named typeface of his own based on his lettering style. Dave Gibbons (the typeface) is infinitely superior to its imitator.

Putting aside that it doesn't respect the traditions of comics, Comic Sans has problems with the consistency of the letterforms and the spacing between them (FIGURE 12.9).

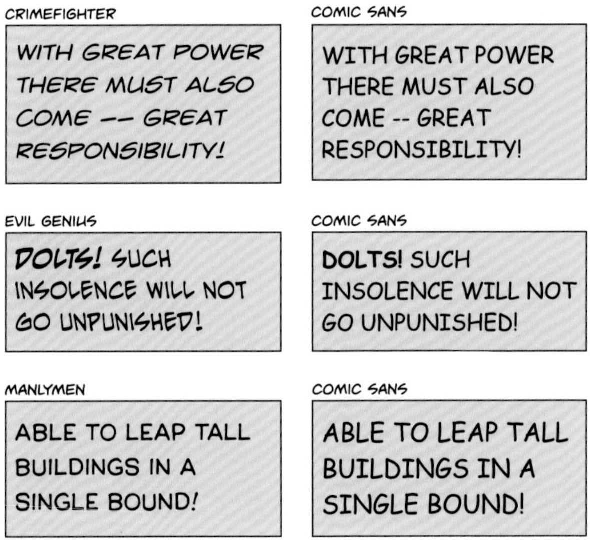

FIGURE 12.8

Professional comics lettering compared to Comic Sans.

CRIMEFIGHTER

WITH GREAT POWER THERE MUST ALSO COME —— GREAT RESPONSIBILITY!

COMIC SANS

WITH GREAT POWER THERE MUST ALSO COME -- GREAT RESPONSIBILITY!

EVIL GENIUS

DOLTS! SUCH INSOLENCE WILL NOT GO UNPUNISHED!

COMIC SANS

DOLTS! SUCH INSOLENCE WILL NOT GO UNPUNISHED!

MANLYMEN

ABLE TO LEAP TALL BUILDINGS IN A SINGLE BOUND!

COMIC SANS

ABLE TO LEAP TALL BUILDINGS IN A SINGLE BOUND!

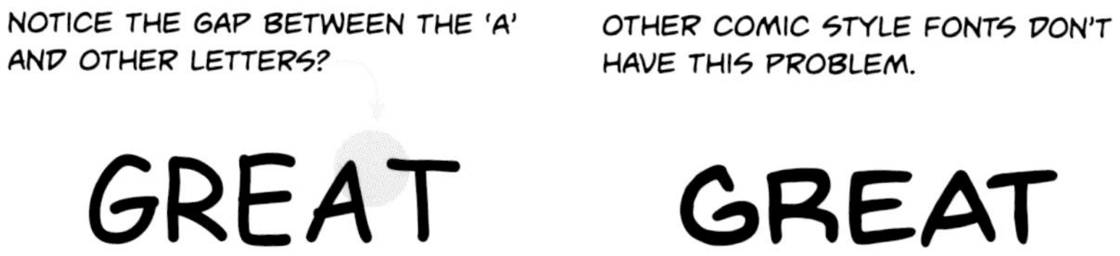

FIGURE 12.9

Kerning problems
in Comic Sans.

GREAT

GREAT

Professional comics letterer Todd Klein's judgment on Comic Sans is univocal:

> Comic Sans fails on every level, and I think deserves the scorn it's gotten. Only the fact that it comes with so many Microsoft products, making it easy for the design deaf to turn to, has kept it prevalent. Computer users beware, Comic Sans is nothing more than a way to label yourself clueless about comics, fonts, and good design. (Klein 2009)

If you want a look inspired by comics, go straight to the source. Companies like Comicraft and Blambot, which letter comics professionally, sell many of the same fonts that they use to letter the utterances of Spider-Man, Batman, and Hellboy.

The moral of the story is not "Don't use Comic Sans." The moral is that a bad choice of typeface makes the conversation about the typeface, not the work.

Professional typographers have strong opinions about many other typefaces. Other common targets of scorn include Papyrus, Brush Script, MT Curlz, Vivaldi, Bradley Hand, and Souvenir (Garfield 2011, Bigman 2012, Gendelman 2013, Butterick 2016, Editorial Team 2017). But many popular typefaces have both fans and detractors simultaneously. In the documentary movie *Helvetica*, the opinion of type designers over this one typeface could not be more divided (Hustwit 2007):

> Erik Spiekermann: Most people who use Helvetica, use it because it's ubiquitous. It's like going to McDonald's instead of thinking about food. Because it's there, it's on every street corner, so let's eat crap because it's on the corner.

> Michael C. Place: For me Helvetica is just this beautiful, timeless thing.

"I've heard that hard-to-read typefaces makes people understand things better." A couple of studies reported that using typefaces that were difficult to read improved performance (Alter *et al.* 2007, Diemand-Yauman *et al.* 2011). These findings made for a great narrative that gained attention in the popular press (Gladwell 2013) – but remember, narrative is like the Force: it has a dark side. Multiple studies tried to replicate this effect unsuccessfully (Thompson *et al.* 2013, Burnham 2015, Meyer *et al.* 2015, Kühl and Eitel 2016). Making something hard to read on purpose is just not helping your readers.

Again, the main thing is that you make conscious decision about the text you are trying to display. A portrait-style poster, for example, might benefit from a condensed font, because narrow letterforms reflect the shape of the page, and it allows more characters for a given width.

Most posters will have at least two levels of text in a visual hierarchy: headings (including the title) and the main text. By using a different typeface, or a different font (e.g., the same style in ultrablack and book weight) for each, you can make it obvious to a reader which parts are which in the poster. Make your two choices look very different (FIGURE 12.10). Contrast between type helps the reader recognize that they are signifying different things (Williams 2004).

FIGURE 12.10

Contrasting heading and body-text typefaces.

WHEN HEADINGS ARE SO SIMILAR TO THE MAIN TEXT, DIFFERENCES LOOKS LIKE MISTAKES!

Chapter II

Now Nio Kuro, a Prince and the most famous hunter in the kingdom, had come in his boat down the river that ran through the haunted wood. With him he had brought many servants and his pack of trained leopards, with which he hunted, and

Chapter II

Now Nio Kuro, a Prince and the most famous hunter in the kingdom, had come in his boat down the river that ran through the haunted wood. With him he had brought many servants and his pack of trained leopards, with which he hunted, and which were swifter and had keener scent than any dogs. Possibly Nio

MAKING HEADINGS AND TEXT *VERY* DIFFERENT LEAVES NO DOUBTS!

CHAPTER II

Now Nio Kuro, a Prince and the most famous hunter in the kingdom, had come in his boat down the river that ran through the haunted wood. With him he had brought many servants and his pack of trained leopards,

Chapter II

Now Nio Kuro, a Prince and the most famous hunter in the kingdom, had come in his boat down the river that ran through the haunted wood. With him he had brought many servants and his pack of trained leopards, with which he hunted, and which were swifter and had keener

A
FONT FAMILY
is designed so that
DRAMATICALLY DIFFERENT TYPEFACES
still work together
harmoniously.

FIGURE 12.11

Font family of
Adorn typefaces.

Because headings are typically a single word, or a few words, you can be a little more daring with the type you choose. You don't have to worry about readability to the same degree as in the main paragraph text.

Example

The type above (Bernhard Modern) is easy to read in a single word and has some style to it. But because of its small x-height, it would be hard reading if you had to wade through entire paragraphs of it.

This is one of the advantages of editing ruthlessly. The more concise you are, the more daring and expressive your typeface selection can be. You can put up with almost any typeface for a word or two.

Many font foundries design font families (**FIGURE 12.11**): a set of fonts for a typeface that are specifically intended to work together (unlike many families of people). By using a font family, you can get variety in your fonts without losing the unity. Some typefaces come in both serif and sans serif versions. A couple of examples are shown in **FIGURE 12.4**.

Placeholder text

Maybe you want to start laying out your posters, but you haven't written the text yet. Maybe you want to see how a new typeface will look. There are lots of little situations where you might want some words for a temporary placeholder.

There's a solution for you. It's called "Lorem ipsum" (see **FIGURE 12.2**). It's somewhat corrupted Latin text (Adams 2001) in which the average word length is roughly the same as in a typical English paragraph, so it makes a good temporary stand-in for text that hasn't been written yet. There are various Lorem ipsum generators online that can give you the amount of text you think you need for a task.

Text spacing

How easily someone can read your text depends on type size, line length, and spacing (Dabner *et al.* 2010), including spacing between the lines, between the words, and between the letters (FIGURE 12.12). But crowded text is a common problem for poster makers still thinking, "Whoever has the most information on the poster, wins." There's not enough room between lines, and not enough room between the letters in each word (Hampton-Smith 2017, Pal 2020). This is made worse by some software (notably PowerPoint) that "autofits" text without warning or asking, by shrinking either the text or the line spacing. If you're not paying attention, your single-spaced text will change without warning to 0.9-spaced text or smaller.

Most people know the difference between single and double spacing, and almost every type of software makes this feature easy to find and change. That space between lines is also called leading (rhymes with "sledding," not "bleeding") by typographers.

Lines of text on posters are often quite long – not in terms of number of words necessarily, but in physical distance. Even if the number of words per line is reasonable, posters can be big enough that when you reach the end of the line you're reading, scanning back to the left to find the next line can be a greater distance than when you're reading a book in

FIGURE 12.12

Changes in the appearance of text with changes in spacing.

ORIGINAL TEXT

And then about him coiled the great, slimy folds of a hideous monster of that prehistoric deep—a mighty serpent of the sea, with fanged jaws, and darting forked tongue, with bulging eyes, and bony protuberances upon head and snout that formed short, stout horns.

REDUCED LINE SPACING

And then about him coiled the great, slimy folds of a hideous monster of that prehistoric deep—a mighty serpent of the sea, with fanged jaws, and darting forked tongue, with bulging eyes, and bony protuberances upon head and snout that formed short, stout horns.

REDUCED WORD SPACING

And then about him coiled the great, slimy folds of a hideous monster of that prehistoric deep—a mighty serpent of the sea, with fanged jaws, and darting forked tongue, with bulging eyes, and bony protuberances upon head and snout that formed short, stout horns.

REDUCED CHARACTER SPACING

And then about him coiled the great, slimy folds of a hideous monster of that prehistoric deep—a mighty serpent of the sea, with fanged jaws, and darting forked tongue, with bulging eyes, and bony protuberances upon head and snout that formed short, stout horns.

your hands. The further back to the left you scan to find the next line, the more likely it is that you will lose your place. Increasing the leading helps make each line distinct in long text.

Single-spaced text often puts lines a little too close together, particularly for a poster. Try increasing the leading to about 115–145% of the point size (Butterick 2016). In most software, you need to find an option that specifies spacing "exactly." You may need more space. Look at the letters in your text. If you have a typeface with very long ascenders (the part that is taller than most lowercase letters, like the pointing-up bits on l, k, and t; see FIGURE 12.3) and descenders (the dropping-down bits on letters like p, g, and y), increase the leading. You don't want your words colliding! If you have a typeface where the lowercase letters are very close to the uppercase letters in height (known as a large x-height), again, you'll want to increase the space between the lines. Letters with a large x-height tend to form swaths of gray if placed too close together.

The space between words is, uh, word spacing. There are two other terms that refer to spacing between letters: tracking and kerning. Tracking is the regular distance between letters in a line, and kerning is spacing made between two specific letters. These are fine-tuning adjustments that will be discussed in Chapter 20.

Do not type two spaces after a period. I know many people have been taught to type two spaces after a period. Many people are surprisingly passionate about this. There is historical justification for the practice (Heraclitus 2011, Bricker 2013), but professional typographers haven't used two spaces after a period for decades. A single space between sentences is the standard for contemporary typesetting. One piece of evidence that this is so is that if you put double spaces in the code for web pages, internet browsers automatically suppress them and show single spaces instead.

Life is a finite number of keystrokes. Don't use any of them unnecessarily.

Text distortion

If you need a narrow font, do not squish a font (Landa *et al.* 2007). Use a condensed font specifically designed to show a lot of letters in a small space (FIGURE 12.13).

While the distorted and condensed text take up the same amount of space, and both let you complete the sentence in the allotted width, comparing the two shows they are subtly different. The condensed text looks a little thicker and a little more regular than the squished text. The longer the phrase, the more this matters.

FIGURE 12.13

Condensed text.

NORMAL TEXT

It was a lens, some four inches

SQUISHED TEXT

It was a lens, some four inches across.

CONDENSED TEXT

It was a lens, some four inches across.

Vertical text

There are few clean ways to stack text, because of the varying letter widths (FIGURE 12.14; Landa *et al.* 2007), so stacking letters on top of one another should be avoided. Usually, the only place you will find vertical text is alongside the y-axis of a graph. There, software packages normally make the decision to rotate the text rather than stacking the letters on top of each other.

Monospaced fonts, which are designed to occupy the same width, can be acceptable. Decorative fonts with a geometric background work even better for this rare situation.

FIGURE 12.14

Vertical text.

Chapter recap

- Typefaces are some of a poster designer's best tools for creating a distinctive look.

- Sans serif fonts generally perform well on posters.

- Setting text too close together and distorting letter forms are common errors in design.

CHAPTER 13

Layout

At this point, you have some basic ideas about your poster. You should know what your title is, and you should know what some of your main figures will be. It's time to start working on laying out the poster.

The size and shape of your poster makes a big difference to layout strategies. This is determined not only by how big the conference's poster boards will be but by how you plan to print the poster. For example, the mounting space of many poster boards is 48 inches (1,220 mm) high, but some plotter printers use rolls of paper that are 36 or 42 inches (915 or 1,065 mm) wide, so you cannot make a poster that is 48 inches high.

Poster real estate

"Location, location, location" is the mantra of real estate. Prime real estate is in the best locations. Real estate on a poster can be divided into three key locations, which are determined by the height of the poster board and the typical height of the viewers (FIGURE 13.1).

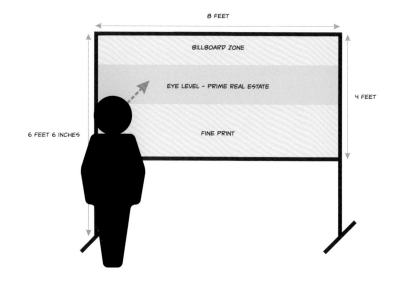

FIGURE 13.1

Eye levels and poster viewing.

A typical poster board is maybe 6½ or 7 feet (200–215 cm) high. That's taller than the usual height of people looking at your poster, which is probably somewhere between 5 and 6 feet (150–185 cm).

Prime real estate on a poster is about eye level, which is around 5–5½ feet (150–170 cm) off the ground. This is the region where people will naturally tend to look when viewing the poster. That's where you want to put your important stuff.

Titles sit above eye level, and above the top of people's heads. Titles are like billboards or highway signs that rise above the background and can be seen by anyone walking around the session hall.

The space below eye level is the least valuable real estate on your poster. It's usually the biggest available space on the poster, but people have to deliberately look down to see what's there, and that space gets concealed by people in front of the poster.

Reading gravity

Reading is hard. It takes years to learn how to read, because there are a lot of rules to internalize. There's not only the shape of letters and the sounds they make but how those words are laid out on a page. People who read English fluently expect text to follow patterns. Audience members at a research poster session are experienced readers – perhaps some of the most fluent and proficient readers in the world – and they know enough to follow conventions established for the written English language. But academics are so comfortable with reading that they seem to forget that those conventions exist when designing a poster.

If you are reading an English edition of this book, you started this page in the upper left-hand corner, because that's the convention. You expect to finish the page in the lower right-hand corner. The technical term for "start upper left, end lower right" is *reading gravity* (FIGURE 13.2). But I call it "the Cosmo rule."

FIGURE 13.2

Reading gravity and viewing zones.

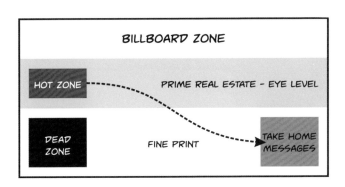

Magazine covers face a lot of the same design challenges as conference posters. They are often being viewed in busy environments where people are distracted, like supermarket checkout lines. The magazine has only a few short moments to convince a potential buyer, "Hey! You! Over there! Yes, you! Come here and read me!"

Cosmopolitan magazine understands this. Every issue of *Cosmo* has a sex story on the cover, because sex sells. Headlines like "Sex you crave – daring moves that feel extra awesome for you," "56 hot sex tips!" and "Heat up sex!" are *Cosmo*'s white rabbits: something to convert a casual passer-by into a buyer and reader. And when you look at where the sex story on the *Cosmo* cover is, it's invariably in the upper left-hand corner. Because that's where people will look first.

Likewise, comics readers look for starting panels in the upper left corner (Pederson and Cohn 2016). Eye tracking of people reading newspapers also shows that people spend longer on articles on the left side of the page, making it a favorable position for important or attention-getting material (Holmberg 2004).

There are two lessons here. The first is a general reminder that your layout should guide people through your poster without resorting to explicit directives as to where to go next. The second is that the upper left space below the title is the most valuable part of your poster besides the title itself, so you'd better put something awesome there. That's the hot zone.

That bottom right corner is a contradiction. On the one hand, it's away from eye level, making it less valuable. But on the other hand, it's the natural "stop" point for a narrative, so people will tend to look in that general area when they are looking for the ending. It's like the final chapter in the murder mystery that people will skip ahead to when they just want to know whodunit. It's a good place to put take-home messages.

The habit of following "reading gravity" means people will tend to look up initially to the hot zone, then drift down towards the bottom right corner. But that pattern is not the only reading rule people have learned. Other habits of reading can direct readers through the poster. There are some layouts that are hard to screw up. And there are some that just never work, no matter how hard you try.

Chapter recap

- The top of a poster is more valuable space than the bottom of the poster.
- People expect to follow a consistent path from the top left of the poster to the bottom right of the poster.

CHAPTER 14

Grids

If there is just one thing that I hope you take away from this book, it is this:

Make a grid.

A grid is a series of lines – almost always horizontal and vertical – that guide the positioning of elements on a page but that are ultimately not printed on a document. Grids have helped organize information for thousands of years. Many of the earliest writing systems – such as cuneiform tablets, Egyptian hieroglyphics, and Chinese characters – line up symbols on a grid, whether horizontal or vertical.

Remember that part of good design is about similarity, and that includes similar width or similar alignment between objects. Nearly every edge of an object on a page should be aligned with the edge of some other element (FIGURE 14.1). The word "edge" is important here, because aligning the center of objects to each other usually does not create the same sense of care and order that aligning edges creates.

FIGURE 14.1

Aligning the centers creates misaligned edges.

ALIGNING THE CENTER OF EACH COLUMN DOESN'T HIDE MISALIGNED EDGES!

The moon, by her comparative proximity, and the constantly varying appearances produced by her several phases, has always occupied a considerable share of the attention of the inhabitants of the earth.

From the time of Thales of Miletus, in the fifth century B.C., down to that of Copernicus in the fifteenth and Tycho Brahe in the sixteenth century A.D., observations have been from time to time carried on with more or less correctness,

until in the present day the altitudes of the lunar mountains have been determined with exactitude. Galileo explained the phenomena of the lunar light produced during certain of her phases by the existence of mountains, to which he assigned a mean altitude of 27,000 feet. After him Hevelius,

an astronomer of Dantzic, reduced the highest elevations to 15,000 feet; but the calculations of Riccioli brought them up again to 21,000 feet.

At the close of the eighteenth century Herschel, armed with a powerful telescope, considerably reduced the preceding

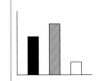

measurements. He assigned a height of 11,400 feet to the maximum elevations, and reduced the mean of the different altitudes to little more than 2,400 feet. But Herschel's calculations were in their turn corrected by the observations of Halley, Nasmyth, Bianchini, Gruithuysen, and others; but it was

reserved for the labors of Boeer and Maedler finally to solve the question. They succeeded in measuring 1,905 different elevations, of which six exceed 15,000 feet, and twenty-two exceed 14,400 feet. The highest summit of all towers to a height of

22,606 feet above the surface of the lunar disc. At the same period the examination of the moon was completed. She appeared completely riddled with craters, and her essentially volcanic character was apparent at each observation. By the absence of refraction in the rays of the planets occulted by her we conclude that she is absolutely devoid of an atmosphere. The absence of air entails the absence of water. It became, therefore, manifest that the Selenites,

to support life under such conditions, must possess a special organization of their own, must differ remarkably

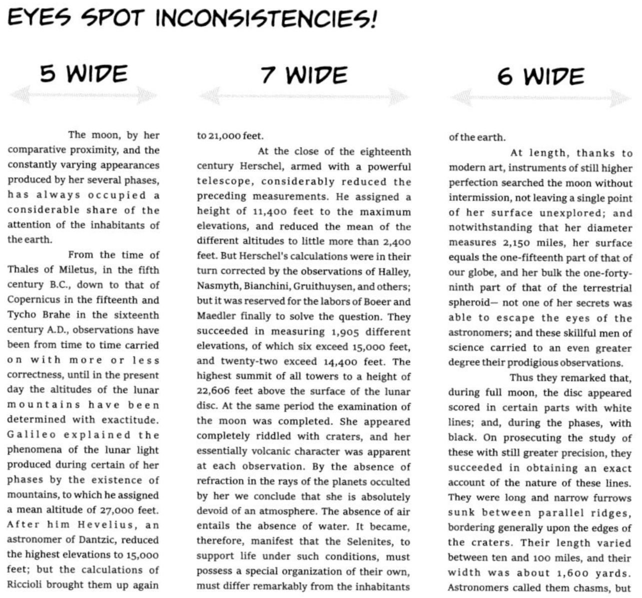

EYES SPOT INCONSISTENCIES!

5 WIDE

7 WIDE

6 WIDE

The moon, by her comparative proximity, and the constantly varying appearances produced by her several phases, has always occupied a considerable share of the attention of the inhabitants of the earth.

From the time of Thales of Miletus, in the fifth century B.C., down to that of Copernicus in the fifteenth and Tycho Brahe in the sixteenth century A.D., observations have been from time to time carried on with more or less correctness, until in the present day the altitudes of the lunar mountains have been determined with exactitude. Galileo explained the phenomena of the lunar light produced during certain of her phases, to which he assigned a mean altitude of 27,000 feet. After him Hevelius, an astronomer of Dantzic, reduced the highest elevations to 15,000 feet; but the calculations of Riccioli brought them up again

to 21,000 feet.

At the close of the eighteenth century Herschel, armed with a powerful telescope, considerably reduced the preceding measurements. He assigned a height of 11,400 feet to the maximum elevations, and reduced the mean of the different altitudes to little more than 2,400 feet. But Herschel's calculations were in their turn corrected by the observations of Halley, Nasmyth, Bianchini, Gruithuysen, and others; but it was reserved for the labors of Boeer and Maedler finally to solve the question. They succeeded in measuring 1,905 different elevations, of which six exceed 15,000 feet, and twenty-two exceed 14,400 feet. The highest summit of all towers to a height of 22,606 feet above the surface of the lunar disc. At the same period the examination of the moon was completed. She appeared completely riddled with craters, and her essentially volcanic character was apparent at each observation. By the absence of refraction in the rays of the planets occulted by her we conclude that she is absolutely devoid of an atmosphere. The absence of air entails the absence of water. It became, therefore, manifest that the Selenites, to support life under such conditions, must possess a special organization of their own, must differ remarkably from the inhabitants

of the earth.

At length, thanks to modern art, instruments of still higher perfection searched the moon without intermission, not leaving a single point of her surface unexplored; and notwithstanding that her diameter measures 2,150 miles, her surface equals the one-fifteenth part of that of our globe, and her bulk the one-forty-ninth part of that of the terrestrial spheroid— not one of her secrets was able to escape the eyes of the astronomers; and these skillful men of science carried to an even greater degree their prodigious observations.

Thus they remarked that, during full moon, the disc appeared scored in certain parts with white lines; and, during the phases, with black. On prosecuting the study of these with still greater precision, they succeeded in obtaining an exact account of the nature of these lines. They were long and narrow furrows sunk between parallel ridges, bordering generally upon the edges of the craters. Their length varied between ten and 100 miles, and their width was about 1,600 yards. Astronomers called them chasms, but

FIGURE 14.2

Inconsistent column widths.

The more elements line up, the more purposeful, composed, and clean your poster looks. A grid is almost incomparably powerful in design because it makes those alignments explicit. It turns messes into successes. It gives you a considered and thoughtful appearance. The tighter the grid, the closer to gods.

Posters with no grids usually look chaotic and amateurish (but see *Breaking the grid*, below).

Often, I find that early decisions about a grid shape a poster most profoundly. Pictures and graphs may move, fonts and colors may change, but the underlying grid often persists throughout. This means that it is important to pick a grid at the start, before laying in content. Too often, I see columns that look as though they have been made arbitrarily to fit some graphic plunked into them (FIGURE 14.2).

Automatic alignment with computers will only take you so far (FIGURE 14.3). Often, you have to fine-tune alignments. I will often just draw quick straight lines as guides to help me see where I want to align things, then delete them later.

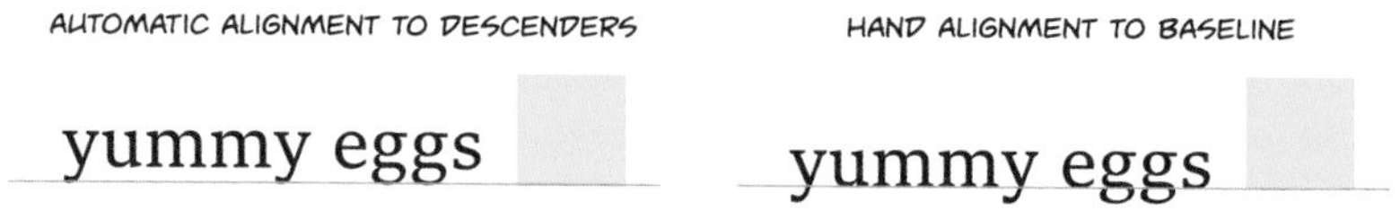

AUTOMATIC ALIGNMENT TO DESCENDERS

yummy eggs

HAND ALIGNMENT TO BASELINE

yummy eggs

FIGURE 14.3

Text alignment to objects.

Remember that design generally, and alignment specifically, is not math. If you divide a page exactly in half along the horizontal, the top half will often look a bit larger than the bottom half (Landa *et al.* 2007). Frame makers know this. The mats in picture or art frames are thicker at the bottom than at the top. Comic-book heroes are often drawn with small heads, so that they are eight and a half heads tall (heroic proportions). Typesetters often set punctuation outside the edges of a grid (Landa *et al.* 2007). So you may have to nudge objects around until they look right.

Columns and rows

Grids are normally a series of rectangles. They can be organized as vertical columns or as horizontal rows. It is usually a good idea to pick one of these two and stick with it. Make the poster consistently columns or consistently rows, rather than flipping between the two. Which to use depends in part on the shape of the poster.

Whether you chose to organize your poster in columns or rows, you must signal to a viewer which you are using. A common cue that people use is to look for an uninterrupted margin that divides the poster (except for the title). See, for example FIGURE 7.11: there are continuous horizontal margins the cut across the poster, but none running vertically.

That most posters are wider than tall (landscape layout) means that the text and graphics will most likely be arranged in a series of columns. But columns can work in almost any page shape, and it's hard to go wrong with them (Chapter 1). Columns are everywhere in things that work in print: newspapers, magazines, websites, and some books. People understand that you start in the upper left, read down the column to the bottom, then go back to the top of the next column to the right.

The number and width of columns is probably the single most important factor in designing a grid for a poster. The number of columns will depend mostly on the width of the poster and the size of the text. If there is any text that is long enough to form a paragraph, the size of the type has a major impact on column width, and therefore column number.

Think about column width not in absolute numbers like inches or millimeters, but in **number of words per line**. A good guideline for paragraph width is about ten to twelve words per line of text (Jury 2002). This is only a rough guideline, however, and there are many examples where it is not followed. Some magazines have narrower columns, maybe six to eight words wide. On a poster, it is probably better to err on the side of narrow columns rather than wide ones (Ogilvy 1963). Reading a wide column of text can be difficult for someone who has limited peripheral vision, or has trouble scanning because it is difficult to move their head from side to side (Parks Canada 1993). While zipping from line to line in a narrow column might be annoying in a very long piece of text, it's not such a big problem for a poster. But the narrower the columns, the more likely it is to create awkward gaps between letters or words, or ragged edges on the right (FIGURE 14.4).

Remember that photographers and graphic artists often talk about the "rule of thirds" (see Chapter 8). Visually interesting images, pages, or photographs often have their focal points off-center, at one of the four points formed by intersections of imaginary lines that divide the space into thirds both vertically and horizontally. Some designers find dividing spaces into halves or quarters too symmetrical and staid, like folded paper. You may want to try for an odd number of columns, like three or five. As a rule of thumb, take the width of your poster in whole feet and round down to the nearest odd number: for example, three columns for a 3- or 4-foot (90–120 cm) poster, five columns for a 6-foot (180 cm) poster.

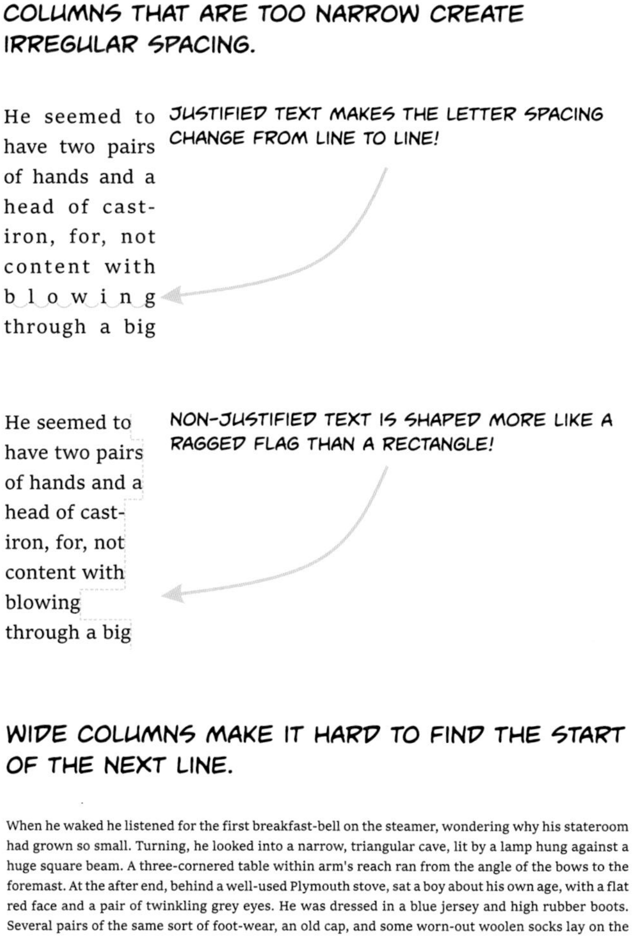

COLUMNS THAT ARE TOO NARROW CREATE IRREGULAR SPACING.

He seemed to have two pairs of hands and a head of cast-iron, for, not content with b l o w i n g through a big

JUSTIFIED TEXT MAKES THE LETTER SPACING CHANGE FROM LINE TO LINE!

He seemed to have two pairs of hands and a head of cast-iron, for, not content with blowing through a big

NON-JUSTIFIED TEXT IS SHAPED MORE LIKE A RAGGED FLAG THAN A RECTANGLE!

FIGURE 14.4

Column width problems.

WIDE COLUMNS MAKE IT HARD TO FIND THE START OF THE NEXT LINE.

When he waked he listened for the first breakfast-bell on the steamer, wondering why his stateroom had grown so small. Turning, he looked into a narrow, triangular cave, lit by a lamp hung against a huge square beam. A three-cornered table within arm's reach ran from the angle of the bows to the foremast. At the after end, behind a well-used Plymouth stove, sat a boy about his own age, with a flat red face and a pair of twinkling grey eyes. He was dressed in a blue jersey and high rubber boots. Several pairs of the same sort of foot-wear, an old cap, and some worn-out woolen socks lay on the

FIGURE 14.5

Six-column layout.

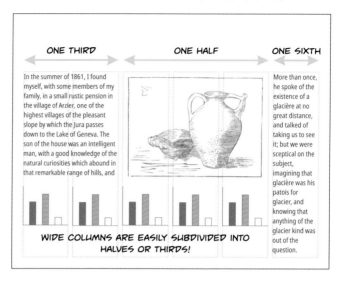

While a three-column grid is a very rugged and forgiving foundation to build a poster from, a six-column grid provides a lot of options (FIGURE 14.5). Think of it as a three-column grid with a secret weapon: you can show two things side by side very easily, without weakening the three-column structure. A six-column grid allows you to vary the width of elements while keeping an orderly structure. The title bar might run across all six columns; a key graph might span three or four columns; text or small graphs might take up one or two columns. But the key is that everything is a multiple of a single column width. While earlier in the book I extolled the virtues of making elements on a poster similar in width, the differences in widths are deliberate, not happenstance. FIGURES 7.6 and 7.13 are examples of posters laid out on a six-column grid.

While you can make different numbers of columns, there will not be many cases where a twelve- or twenty-column grid will be helpful. The individual page elements will get too small at a single column width. Too many columns will start to be indistinguishable from having no grid at all.

Posters that are taller than wide (portrait orientation) might have grids that divide the space into rows. Rows follow our habit of reading from left to right, but they tend not to work so well when text is involved, because it's difficult to create a row of unbroken text that's readable (FIGURE 14.6).

A row will normally be subdivided into smaller sections, which can cause confusion as to how to read the poster. In layouts like those shown in FIGURE 14.7, it's not clear whether the poster is meant to be read across in columns, or down in rows. "Two wides and a tall" or "blockage" creates an ambiguous reading order that confuses readers (Cohn 2013).

INSTEAD OF THIS HARD-TO-READ *WIDE* COLUMN...

Kpfwuvta 4.0 ku vtcpuhqtokpi vjg yqtnf qh ocpwhcevwtkpi cpf yknn dg vjg oquv rqygthwn ftkxgt qh kppqxcvkqp ykvjkp vjg ugevqt qxgt vjg pgzv fgecfg1. Cfxcpegf vgejpqnqikgu nkmg eadgt-rjaukecn rtqfwevkqp uaruvgou (erru), ukowncvkqp, xktvwcn cpf cwiogpvgf tgcnkva, cwvqpqoqwu tqdqvu cpf vjg kpvgtpgv qh vjkpiu (kqv) ctg kpeqtrqtcvgf kpvq vjg rtqfwevkqp xcnwg ejckp vq etgcvg c ugconguu, kpvgnnikgpv cpf cikng pgvyqtm vjcv gpjcpego rtqfwevkxva2,3. Vjgug vgejpqnqikgu cpf rtqeguugu vcmg rnceg ykvjkp vjg uoctv hcevqta, yjkej ku, eqpvgzv-cyctg cpf cuukuvu rgqrng cpf ocejkpgu kp ecttakpi qwv vjgkt vcumu4. Vjg kpfwuvta 4.0 kppqxcvkqp jcu yqtmhqteg kornkecvkqpu vjcv ctg dgkpi kfgpvkhkgf dwv vjgug, fwg vq vjg ncem qh tgugctej, ctg ogtgna urgewncvkxg5. Vjg wtigpv swguvkqp vjcv

WHY NOT TWO EASY-TO-READ *SKINNY* COLUMNS?

Kpfwuvta 4.0 ku vtcpuhqtokpi vjg yqtnf qh ocpwhcevwtkpi cpf yknn dg vjg oquv rqygthwn ftkxgt qh kppqxcvkqp ykvjkp vjg ugevqt qxgt vjg pgzv fgecfg1. Cfxcpegf vgejpqnqikgu nkmg eadgt-rjaukecn rtqfwevkqp uaruvgou (erru), ukowncvkqp, xktvwcn cpf cwiogpvgf tgcnkva, cwvqpqoqwu tqdqvu cpf vjg

Vjgug vgejpqnqikgu cpf rtqeguugu vcmg rnceg ykvjkp vjg uoctv hcevqta, yjkej ku, eqpvgzv-cyctg cpf cuukuvu rgqrng cpf ocejkpgu kp ecttakpi qwv vjgkt vcumu4. Vjg kpfwuvta 4.0 kppqxcvkqp jcu yqtmhqteg kornkecvkqpu vjcv ctg dgkpi kfgpvkhkgf dwv vjgug, fwg vq vjg ncem qh tgugctej, ctg ogtgna urgewncvkxg5. Vjg wtigpv

LAYOUTS THAT *NEVER* WORK

TWO WIDES AND A TALL

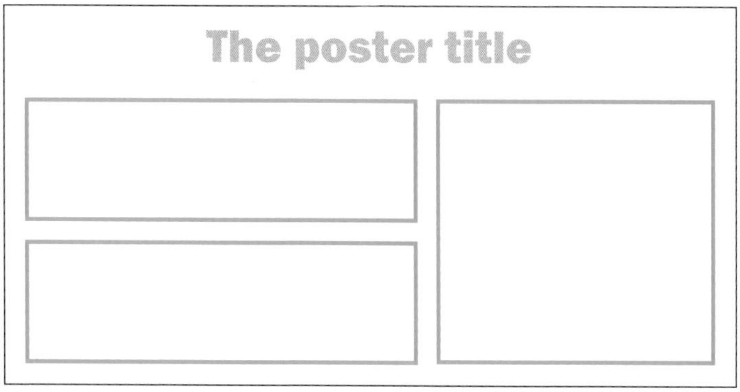

THE SWEDISH FLAG

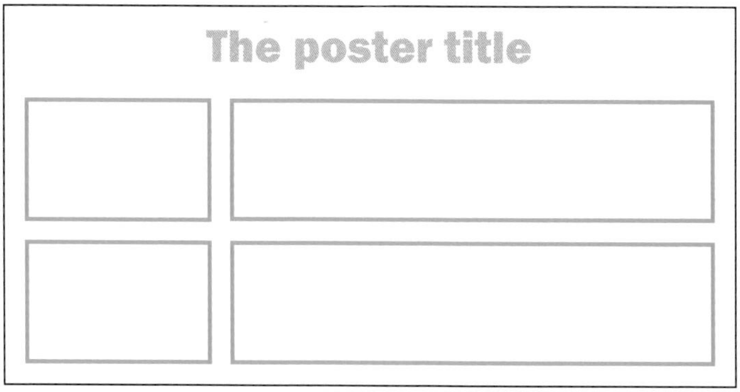

FIGURE 14.6

Wide text columns can be replaced by narrow text columns.

FIGURE 14.7

Layout fails.

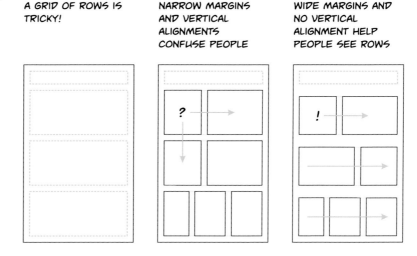

A GRID OF ROWS IS TRICKY!

NARROW MARGINS AND VERTICAL ALIGNMENTS CONFUSE PEOPLE

WIDE MARGINS AND NO VERTICAL ALIGNMENT HELP PEOPLE SEE ROWS

If you plan a grid correctly and think about reading order, people will follow the order intuitively (FIGURE 14.8). Signs, magazines, newspapers, movie credits, comics, and blogs get along without directives to show the next step in the sequence. Let's dig into comics as an example.

Integrating text and graphics

Many young readers start reading with picture books and comics. This isn't just a training device for novice readers. Readers understand the relationships between information presented in text and graphics better when the two are presented together rather than separately, and this prolongs reading, even in experienced readers (Holsanova *et al.* 2005). Like conference posters, comics are a mix of pictures and text, and it's

FIGURE 14.9

Two potential ways to read comic panels.

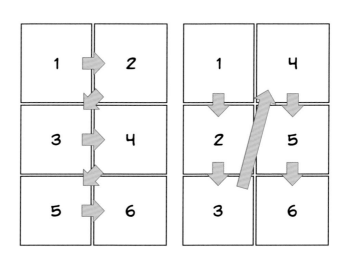

critical to understanding the story to know what order to read in, which means that poster makers would do well to study the conventions and layout in comics. How do comics signify the order of reading?

Comics, like many posters, often have individual panels that are laid out on a grid. Sometimes these grids are regular, but sometimes not. Panel layouts can be complicated. And even simple grids have potential ambiguities in reading order, as shown in FIGURE 14.9. Many Silver Age comics from the 1960s are laid out in this way, with two squares across and three squares down.

White space is typically used to organize the panels, showing which elements are grouped together, but in a simple grid like this the white space alone does not provide a strong enough cue. FIGURE 14.10 shows the panel layout and word balloon placement from a classic comic-book page

FIGURE 14.10

Word balloon placement from classic comic page.

FIGURE 14.11

Poster signposting
the reading order
using numbers
in the center.
(Jacquelyn Gill)

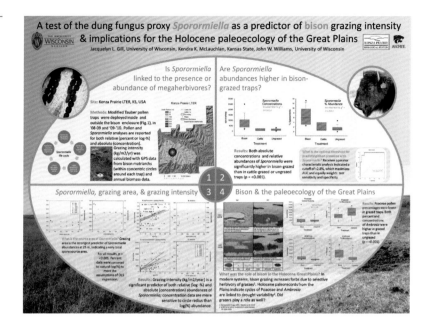

drawn by Jack Kirby and lettered by Artie Simek, and shows how critical
the placement of word balloons is to reading in the correct sequence.

When you're reading the first panel in the upper left corner, the next
closest balloon is to the right, not down, thus encouraging the correct
reading sequence. The word balloons run almost straight across the top
of each panel, close to each other. The artwork underneath distances the
balloons from each other in the vertical dimension, making a reader less
likely to wrongly follow down first instead of to the right. The moral of
the story is to put related objects close to each other (proximity again:
see *Visual hierarchy* in Chapter 7).

In a poster, well-defined white space can serve the purpose that art does
in comics. Clear bands of uninterrupted white space can be strong cues
to the reader as to the direction of the reading flow.

If you ever do end up with an unusual or ambiguous reading sequence,
it is useful to signpost this to make the reading order explicit, as in
FIGURE 14.11. Signal your intentions!

Breaking the grid

Grids create order, but they can also create monotony. Perfect symmetry
is boring. Again, experienced photographers use the "rule of thirds" and
don't place their focal point in the center of the frame. If you watch a
movie or television, look at where important objects, like actors' faces in

close-up, are placed on the screen. They are usually not in the middle of the screen. They are usually about a third or two-thirds of the way across the screen. Comics have transitioned away from grids to more layouts that consider whole pages (Pederson and Cohn 2016) but the move away from grids took decades of artistic innovation (McGovern 2014) and shifts in reading norms.

Grids are guides, not tyrants. There are ways to ignore the grid – some subtle, some not so.

One way is to run images right through the margin and off the page (Seddon and Waterhouse 2009). Running something over the edge of the printed paper is a "bleed." An element that runs over all four edges is a "full bleed."

Most software works on the assumption that text and images will be rectangular. Rectangles create regularity, solidity, and predictability. But predictable can become boring, so sometimes it is worth thinking about using shapes besides rectangles (FIGURE 14.12). Sometimes this is simply a matter of looking for different shapes within a picture or a graph, or different alignments of elements within a picture.

You can get striking results by building shapes into the poster layout, incorporating them into your grid. More powerful graphics programs can wrap around irregular shapes, or create text to fit in any arbitrary shape.

Many non-technical posters, like movie and advertising posters, use diagonal elements (Lupton 2015). Diagonals can bring action and movement to the poster (FIGURE 14.13). If you want diagonals, build them into your grid, rather than slapping an element on the page and rotating it.

Kimberly Elam (2004) said, "The human eyes loves the circle and embraces it." When I do word-find puzzles, I always look for words with

THE LEFT ICON IS EASIER TO FIT INTO A PAGE, BUT THE RIGHT IS MORE DYNAMIC!

FIGURE 14.12

Flat and diagonal icons.

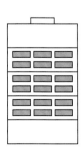
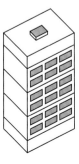

FIGURE 14.13

Sample diagonal
poster layout.

Let me say a few words on these two classes of experiments,--Experiments of Illustration and Experiments of Research. The aim of an experiment of illustration is to throw light upon some scientific idea so that the student may be enabled to grasp it. The circumstances of the experiment are so arranged that the phenomenon which we wish to observe or to exhibit is brought into prominence, instead of being obscured and entangled among other phenomena, as it is when it occurs in the ordinary course of nature. To exhibit illustrative experiments, to

encourage others to make them, and to cultivate in every way the ideas on which they throw light, forms an important part of our duty. The simpler the materials of an illustrative experiment, and the more familiar they are to the student, the more thoroughly is he likely to acquire the idea which it is meant to illustrate. The educational value of such experiments is often inversely proportional to the complexity of the apparatus. The student who uses home-made apparatus, which is always going wrong, often learns more than one who has the use of carefully adjusted instruments, to which he is apt to trust, and which he dares not take to pieces.

It is very necessary that those who are trying to learn from books the facts of physical science should be enabled by the help of a few illustrative experiments to recognise these facts when they meet with them out of doors. Science appears to us with a very different aspect after we have found out that it is not in lecture rooms only, and by means of the electric light projected on a screen, that we may witness physical phenomena, but that we may find illustrations of the highest doctrines of science in games and gymnastics, in travelling by land and by water, in storms

an O first, because they are the easiest to spot in a jumble of letters. This may be why bullet points are popular: they provide circles in a place that has few. Considering that posters are built around rectangles, using a circle in the layout is an attention grabber (FIGURE 14.14).

Some posters use a large central circle or oval as a grid, then divide the circle into quarters (FIGURE 14.15).

The circle makes it different and appealing, but it is tricky to make it work. First, you must work out in your software how to make the text fill a non-standard space. Second, even if you can make text wrap well, other rectangular elements (traditional bar graphs, for example) might not fit into the space so neatly. Third, a circle loses some of the cues for normal reading order, so you must make it clear to readers what order to read the sections in.

FIGURE 14.14

Circular poster
layout.

CIRCULAR LAYOUTS STAND APART FROM THE CROWD, BUT NEED *LOTS* OF THOUGHT PUT INTO THEM!

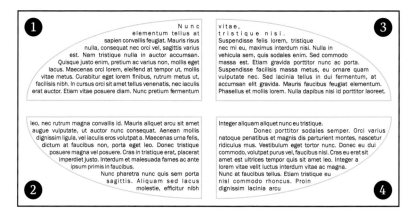

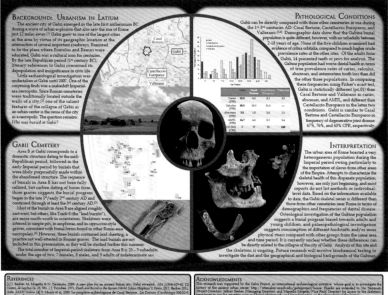

FIGURE 14.15

Poster using circles.
(Kristina Killgrove)

You can use circles in more modest ways, such as highlighting key information on graphs or text.

The examples above are still grids. They are just not *rectangular* grids.

The importance of grids has fluctuated throughout history (Landa *et al.* 2007). The organic styles of Art Nouveau submerged grids, while the later geometric styles of De Stijl emphasized grids. But the disregard for rectangular grids finds its purest expression in "grunge" typography. Grunge typography is a style that takes its aesthetic from spontaneous art, like graffiti. Grunge typography revels in the unfettered and spontaneous. Designer David Carson is highly influential in this space, and his approach is almost diametrically opposed to that of most of this book (Lanks 2017). Carson sees grids as mostly unnecessary and leading to "boring" design. He prefers to lay out material by eye, emphasizing active decision making rather than computer-assisted alignment. Carson's designs have been criticized for being hard to read, but he considers this "way overstated." "People read what they're interested in reading" (Lanks 2017).

Removing grids and other forms of guidance is a legitimate choice, but it is a difficult choice to pull off successfully. Intense variety can cause disorder, chaos, and confusion (Landa *et al.* 2007). It is possible to unify

FIGURE 14.16

Grunge-style poster.
(Martin Rolfs)

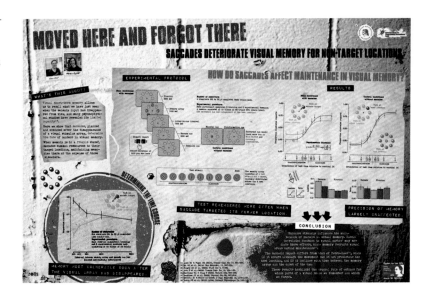

that variation, but doing so relies on an ability to see whole pages and strike balances that create an appropriate response in viewers. Because this is an approach that is more interested in the emotive aspects of a design, it veers more closely to artistic self-expression than most graphic design. But the existence of successful designs without regard to many of the niceties in this book is a useful reminder that there are many ways to create successful designs.

Martin Rolfs' poster in FIGURE 14.16 successfully incorporates some elements of grunge into its design. It doesn't use a grid, other than for paragraphs of text. There is texture and splatter. Edges are ripped. But the overall effect works.

But in most cases the needs of an audience member in a poster session might not be well served by such an anarchic style.

White space and margins

One of the most undervalued things on a poster is nothing. Margins. People underestimate how much white space is on most documents. Just the margins on a standard letter-sized (or A4) piece of paper typically occupy 37% of the page.

If your poster is in evenly divided columns, make your margins at least 15% of your column width (Parks Canada 1993). For example, for a 12-inch (30 cm) column, margins should be at least 1¾ inches (4.5 cm).

One of the most common problems of posters – even ones that are well organized – is there is not enough to space between the elements of

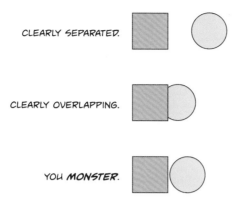

CLEARLY SEPARATED.

CLEARLY OVERLAPPING.

YOU *MONSTER*.

FIGURE 14.17

Tenuous touch between graphic elements.

the poster, and everything looks cluttered. Having objects that are close but not touching creates an uncomfortable visual tension (**FIGURE 14.17**). Separate things. Or put them together. Don't have them almost touching.

White space is a powerful way to organize the elements of a poster and define relationships (Pederson and Cohn 2016). Organize white space as much as you organize information (Williams 2004). White space is usually a better way to separate elements than boxes. Boxes are often a desperate attempt to control an out-of-control layout. Even problematic layouts, like the Swedish flag layout (**FIGURE 14.7**), can be made tolerable by using enough space to signal whether the contents are meant to be read across or down.

Typically, try to leave at least an inch (2.5 cm) between major elements. More space, like maybe 2 inches, is even better.

Watch out for "trapped" white space (Williams 2004): caught between things where it's not necessary, like a heading and body text.

Boxes and junk drawers

When people are laying out posters, they are often so driven to find ways of putting on more stuff that they will cut back on the white space between elements. They know intuitively that the poster has become harder to read because there is so much stuff. But rather than making the tough editorial decision to reclaim the white space, they start putting things in boxes in a desperate attempt to impose some order on the chaos.

Although white space almost always works better than lines and boxes to organize elements, boxes can sometimes be an effective design element. If you do use boxes, boxes with thin lines almost always look better than those with thick lines. Boxes that are defined by a light background color with no borders can work well, too.

FIGURE 14.18

Text box with
rounded corners.

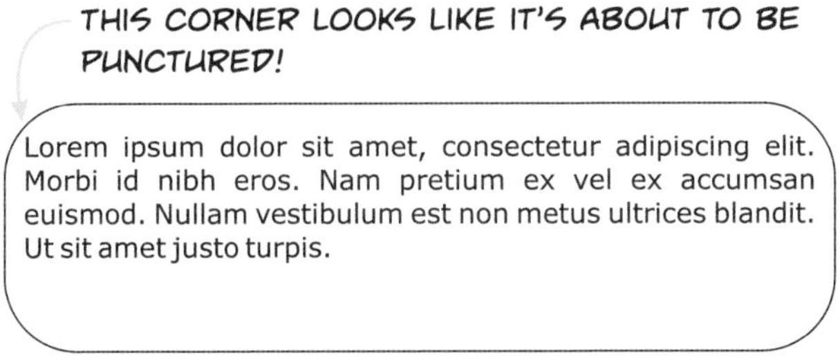

THIS CORNER LOOKS LIKE IT'S ABOUT TO BE
PUNCTURED!

Lorem ipsum dolor sit amet, consectetur adipiscing elit.
Morbi id nibh eros. Nam pretium ex vel ex accumsan
euismod. Nullam vestibulum est non metus ultrices blandit.
Ut sit amet justo turpis.

Boxes with rounded corners seem like an easy way to break the monotony
of rectangles. But there are two problems that make round-cornered boxes
tricky to pull off. First, most elements on a poster tend to be rectangular.
Text blocks are rectangular (unless you labor mightily). Photographs are
rectangular by default. So are most graphs and plots. Put any of those in
a round-cornered box, and the corners of the object within will approach
the round edge of the box. The effect looks like a balloon about to be
popped: a perfect example of the tension created by elements that are
almost but not quite touching (FIGURE 14.18).

To make matters even more complicated, the degree of "roundedness" on
box corners is not something that software does well automatically. Two
rectangles that have the same settings for their corners may be an visual
mismatch, particularly if the rectangles are different shapes (FIGURE 14.19).

Using boxes on posters often leads to the "junk drawer" syndrome. People
treat each box as a big container, and while the boxes themselves may be
aligned, the contents within the box are disorganized and chaotic and
unaligned. Material within boxes should be as carefully aligned as the
boxes themselves.

FIGURE 14.19

Corner rounding
varies with box size.

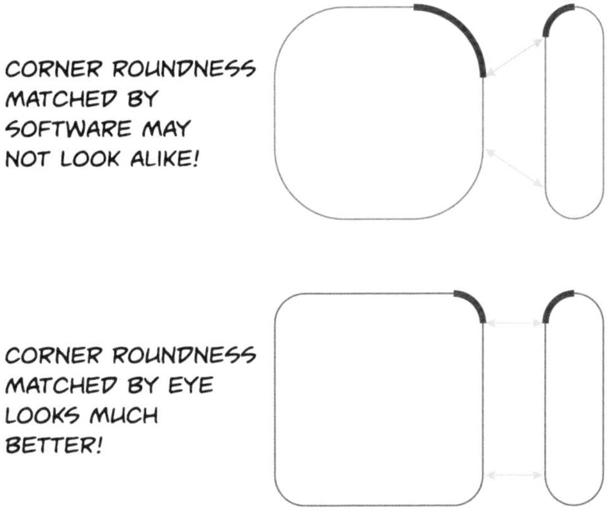

CORNER ROUNDNESS
MATCHED BY
SOFTWARE MAY
NOT LOOK ALIKE!

CORNER ROUNDNESS
MATCHED BY EYE
LOOKS MUCH
BETTER!

Cropping

Many elements on a poster may need cropping to align to a grid. Software often adds margins – which may be helpful if you are printing just that one item but are not needed when it is just one piece of the layout you are working on (FIGURE 14.20).

If you have photographs, you may have to crop them to make them fit into constrained spaces. This can give you a chance of a better composition than the original photograph had. If the subject of the original photograph was centered, you might crop it to follow the rule of thirds.

You can also crop pictures by having them run through the margins and off the page entirely, which is called a "bleed" (see *Breaking the grid*, above).

THIS EXPORTED GRAPH HAS MARGINS YOU DON'T NEED!

WHAT YOU'RE GIVEN

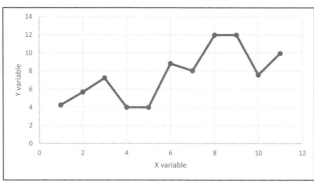

WHAT YOU WANT

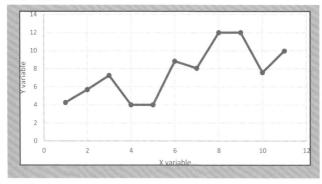

FIGURE 14.20

Excess graph margins.

Chapter recap

- Make a grid and line up the edge of every element on the poster to at least one other element.

- White space is not wasted space. White space organizes your content.

- Boxes usually look desperate rather than organized.

CHAPTER 15

Background

You're almost ready to start putting things on your poster. Don't worry if you haven't made every decision yet – the decisions may change as you go, and that's okay.

Most documents have a few core elements that they share, like words and pictures. But one is so basic and absolutely universal that it's easy to overlook: the background, or the blank page, that you'll be working on.

Plain backgrounds

To start with, leave your background white. In fact, make everything on your poster black, white, or gray to start with (Williams 2004). If it works in monochrome, you know you have a solid format for your poster.

Monochrome does not mean boring. We don't see in black and white, so black and white images can be interesting. The masters of black and white photography make images that are anything but boring. But even if black and white is less interesting than color, that's not necessarily a bad thing. Think about your favorite book. The way it is laid out on the inside is probably boring. There is a reason that most books are printed with black text on a white background: it just works.

Instead of white, you might make the color slightly off-white or cream. These colors mimic the look of paper in many books, which are not always perfectly white.

If you use a dark background, or black, make the letter forms a little heavier than normal. Screens glow but ink flows, so letters may not look as big as you expect from a computer screen.

Photo backgrounds and overlap

Many people, fearing that black on white is boring, will create a photo backdrop for the whole poster, thinking it is "eye-catching." This is often a disaster, a lesson I learned from watching the movie *Tora! Tora! Tora!* The film dramatizes the attack on Pearl Harbor. Actors often spoke Japanese, and the filmmakers subtitled the Japanese dialogue in white type. But many characters in the movie wore white naval uniforms. Entire words and phrases became almost impossible to read, turning the movie-going experience into a frustrating task of guessing missing words.

The moral of this story is that you should take great care if you plan to overlap elements on a poster. Because photos have light and dark areas, it's difficult for a photo to be placed so that it doesn't obscure the text. And if you have made the photo light enough to be able to read the text, you might have made it so light that you can't make out what the photo is showing. For overlap to work well, there needs to be enough contrast between the overlying figure and the background so that the two can be consistently distinguished.

If you do use a photograph of some sort for a background, sometimes it can be used to help organize the content. An area that is mostly white or mostly black might be a logical place to put content, rather than having it straddle areas of different colors.

Placing text directly on top of photographs can be an effective method for labeling or describing images. If the image has a lot of detail in the background that makes distinguishing the text on top difficult, create an even background for the text to sit on. A good strategy is to blur the area of the photograph where the text will go. The box holding the text can be transparent, to minimize the difference between the part of the photograph with and without the text.

Be aware of clichéd backgrounds too. If you are making a poster about astronomy or other space sciences, trust me, you are not the first person to use a background of a night sky or starfield.

Chapter recap

- A busy background such as a photograph obscures content.

CHAPTER 16

Title bars

Most posters devote a large strip at the top to a title and a few other elements. People occasionally try placing titles in other places, such as the middle or the right. These can be interesting but suffer from one problem: they can be blocked by people walking in front of them. Unless your conference has an abnormally large number of abnormally large people, putting the title at the top makes it visible above the heads of other viewers.

With lots of people walking around, the lower part of the poster may be obscured, so the poster title must be high and large to be seen by as many people from as far away as possible.

Titles

If all has gone according to plan, your title should not have changed since you submitted your abstract. But if you need to change it, change it now. It will not matter much if the title on the poster is a bit different than the title in the abstract.

As mentioned in Chapter 5 (see *Writing the title*), the title is the single most important thing on your poster. The layout of the title on the poster should reflect that. Nothing should compete with the title. When someone is walking through a conference, most entries are irrelevant to them, so they skim the titles. Make your title visible and memorable.

Conference organizers often recommend that poster titles be about 72 points when printed. If anything, this recommendation is too conservative. Parks and museums recommend text be 148 points for exhibits viewed from 10 feet (3 meters) away (Smithsonian Accessibility Program n.d., Parks Canada 1993, Harpers Ferry Center Accessibility Committee 2017). The number of people attending the conference makes a difference to your design. Bigger conferences mean that the walkways will be wider and people will be further away, which in turn means that your title may need to be even larger to ensure people can read it. Your title should look almost stupidly big when you are looking at the design on the computer screen.

Mike Morrison (2019) takes this to the extreme. He advocates a style where most space is devoted to a single statement, surrounded by lots of negative space for emphasis (FIGURE 16.1). We already touched on the notion that good titles are short statements of main points. There are a couple of downsides to this format. First, if the conference hall is even slightly busy or crowded, people block the view of the poster's main message, because it is closer to the middle of the poster than the top. Second, increasing the amount of space for the central message means there is less space for everything else. It takes heroic message discipline to keep only the essential material at a readable size instead of simply shrinking everything.

Many people set their poster titles entirely in capital letters (uppercase). There are a few problems with this.

First, using only uppercase letters is the typographic equivalent of shouting at the top of your lungs.

Second, words that are in uppercase are harder to read. When I look at many titles in capitals, I can feel my brain "gearing down" every time. Reading slows to a crawl. The experience frustrates me intensely. This is partly because uppercase letters are all the same height, whereas lowercase letters have a mix of letters with ascenders, descenders, or neither of those. The words make more distinct shapes (FIGURE 16.2).

Headline casing (where every major word starts with a capital) is less bad, but we don't read sentences like that. And it can cause confusion when capitalization is a standard part of formatting (as for species names in biology, where the second part of the name is in lowercase letters).

FIGURE 16.1

Billboard-
style poster.
(Milan Klöwer)

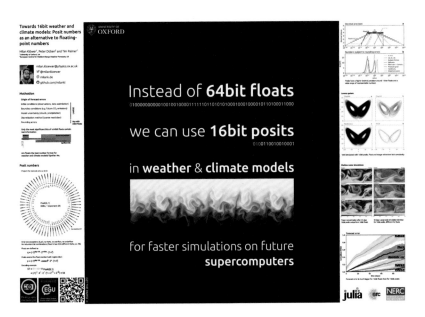

BETTER POSTERS

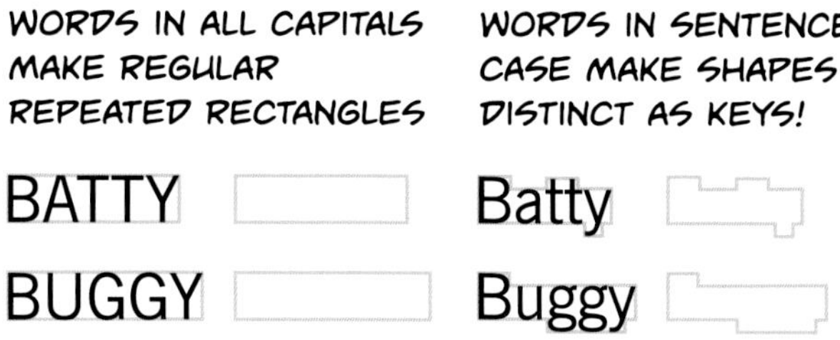

WORDS IN ALL CAPITALS
MAKE REGULAR
REPEATED RECTANGLES

WORDS IN SENTENCE
CASE MAKE SHAPES AS
DISTINCT AS KEYS!

BATTY

Batty

BUGGY

Buggy

FIGURE 16.2

Word shapes created using all capital letters.

Making the title big is necessary for visibility from a distance. The title is also going to sit at the top of your poster, up in the billboard zone. Those two cues alone put the title clearly at the top of your poster's visual hierarchy. You probably don't need to use more emphasis. If you are clear, you don't have to be loud and use all capitals for SHOUTY EMPHASIS.

Most posters have their titles centered, but people may not be able to give you a clear reason why. There is no advantage in reading. Most word processors and other publishing programs start with text left-aligned by default, not centered, which suggests that people are centering their titles deliberately.

Centering text is mostly a holdover from the days before computers. Before computers were used for layout, typography was a tricky, difficult craft. Words had to be set by physically inserting small blocks of metal or wood. Centering words, with varying letter widths and sizes, by moving blocks of type around on the page one by one was complicated, time-consuming, and therefore expensive. Centering became a signal for classy and dignified. Now, it's as easy to center text as not, but the cultural cue has remained. David Jury (2006) wrote, "Centred arrangements were, and still are, considered to be the appropriate way of presenting a text of distinction."

Because we associate "centering" with "expensive" even though there is no expense at all for doing so, centering text has tended to be overused. If you look at a bulletin board or message board, the notices people have posted on it may be on different colors of paper and use a dizzying array of typefaces, but almost all of them will probably have the text centered on the page.

You may not want this level of gravitas on your poster. Centered text can make a document look like a gravestone (Lupton 2004). But the point is not, "Don't center the title of your poster." The point is not to adhere to an out-of-date convention that people follow without ever noticing. Again, conscious decision making is core to design.

There are advantages to aligning the title with the left edge. It moves the eye closer to the "hot zone" and follows reading gravity (Chapter 13). It creates a single white space on the right side (if any) rather than one on each side, which reduces the temptation to stuff irrelevant material into the poster corners.

Authors and affiliations

People usually put the names of authors and their institutional affiliations under the title of the poster. Because of the proximity of the two, the author list is at risk of competing with the all-important title. Your name, your coauthors, and your institution may matter a lot to you. But those are typically not the most interesting thing for most people reading a poster. The title is the most important thing.

The block containing the names of authors and institutions need to be far down the visual hierarchy. But the trend in science is towards projects having more authors (Shapiro *et al.* 1994, Duffy 2017, Barlow *et al.* 2018), with some journal articles listing thousands of authors (Castelvecchi 2015, Woolson 2015). As far as I know, no poster has listed a thousand authors, but even author numbers in the double digits would be a challenge to fit into the limited space available on a poster, as shown in FIGURE 16.3.

You can save some space by using the initials for contributors' given names (e.g., K.W. Jones instead of Kevin W. Jones). For that matter, you'd be surprised how much space you can save if you don't use periods after initials (e.g., KW instead of K.W.).

If we think about the needs of the reader for a second, what are the things conference goers might want to know? They certainly want to know who they might be talking to, that is, the poster presenter. They might also want to know the person behind the project, who is usually the most senior professor or staffer, and often the most recognizable "name" the poster might have. But they might not need to know the name of every contributor. With very many authors, the best solution may be to name only a couple of the key authors in the area under the title. The rest of the authors might be written in small text down in the fine print in the corner. There might be a note, "For the full list of contributors, visit this web page," followed by a link address. While this might feel harsh for the contributors, grit your teeth and remember that design is not done for your benefit, or for your friends. It's about what the audience needs.

Even with a reasonable number of authors, most people also want to show the affiliations of those authors, which adds even more space. Again, make decisions about what to show based on what the audience needs (FIGURE 16.3). Readers may not need to know the street address or zip code of the institutions. They may not even need to know the department people are in.

A COMPLEX AUTHOR LIST

Cameron B. Harris[1], Jonathan M. Heckman[1], Heather L. Holderness[1], Nicole A. Howey[2], Dontae A. Jacobs[3], Elizabeth S. Jewell[2], Maria Kaisler[2], Elizabeth A. Karaska[2], James L. Kehoe[2], Hannah C. Koaches[2], Jessica Koehler[2], Dana Koenig[1], Alexander J. Kujawski[1], Jordan E. Kus[1], Jennifer A. Lammers[1], Rachel R. Leads[1], Emily C. Leatherman[1], Rachel N. Lippert[3], Gregory S. Messenger[1], Adam T. Morrow[3], Victoria Newcomb[3], Haley J. Plasman[3], Stephanie J. Potocny[3], Michelle K. Powers[3], Rachel M. Reem[3], Jonathan P. Rennhack[3], Katherine R. Reynolds[3], Lyndsey A. Reynolds[3], Dong K. Rhee[3], Allyson B. Rivard[3], Adam J. Ronk[3], Meghan B. Rooney[3], Lainey S. Rubin[3], Luke R. Salbert[3], Rasleen K. Saluja[3], Taylor Schauder[3], Allison R. Schneiter[3], Robert W. Schulz[3], Karl E. Smith[3], Sarah Spencer[3], Bryant R. Swanson[3], Melissa A. Tache[4]

FIGURE 16.3

Author lists.

THE SAME AUTHOR LIST, STREAMLINED!

CB Harris, JM Heckman, HL Holderness, D Koenig, AJ Kujawski, JE Kus, JA Lammers, RR Leads, EC Leatherman, GS Messenger. *Memory University*.

NA Howey, ES Jewell, MKaisler, EA Karaska, JL Kehoe, HC Koaches, JKoehler. *The University of Southern South Dakota*.

RN Lippert, AT Morrow, V Newcomb, HJ Plasman, SJ Potocny, MK Powers, RM Reem, JP Rennhack, KR Reynolds, LA Reynolds, DK Rhee, AB Rivard, AJ Ronk, MB Rooney, LS Rubin, LR Salbert, RK Saluja, T Schauder, AR Schneiter, RW Schulz, KE Smith, S Spencer, BR Swanson, DA Jacobs. *Pincher Creek College*.

MA Tache, *The University of Texas International*

AN AUTHOR LIST MADE FOR VIEWERS!

Cameron B. Harris, Memory University, poster presenter

Melissa A. Tache, The University of Texas International, principal investigator

Full author list: http://www.uniwebsite.edu

For various reasons, people may not want to group all the authors from a single institution together in the author listing. Many people use superscripts after the names to let people know which person comes from which institution. But this gets very complicated and difficult to read. There is a good case for grouping all contributors from a single institution together. Let "relative contribution" or "alphabetical order" or "whatever other reason you have for deciding the order of authors" be damned.

Logos

One of the most common ways that people undercut their poster titles is by putting institutional logos in their title banner area. Because people often center titles, this tends to leave the two corners empty, and people feel the compulsive need to fill them up. But corners *don't mind being empty* (Williams 2004).

Most people care about your university's logo on your poster as much as they care about the logos on racing cars: not at all. But while logos on a race car do not interfere with the car's ability to drive, logos on a poster can interfere with the poster's ability to communicate. "The logo won't help make a sell or make a point, but the clutter it brings does add unnecessary noise and makes the presentation visuals look like a commercial. And people hate commercials or being sold to" (Reynolds 2007). Most people list their institution next to their name under the title, so a logo adds very little (Reynolds 2011, Purrington 2019). The space you have is limited, and the title should take up as much of that space as possible. Adding logos tends to weaken the visual hierarchy (FIGURE 16.4).

You might want to use a logo if you are in a large program, department, or institution that has had many people go through it, and you want to connect with those alumni and veterans who have moved on. Recognizing the logo makes them feel like part of your "inner group" (see *Talking to strangers* in Chapter 4), and they may come talk to you. This can be a good way to get gossip material. But more often, institutional logos are like in-jokes: they are recognized mainly by a select few who have had a common experience at an institution. They do nothing for most people except to remind them that they are excluded from the joke.

In most cases, the logo doesn't even add recognizable branding. If you look at many university logos, the sameness is palpable. When you're faced with tens, hundreds, or even thousands of university logos at a conference, your logo is not as distinctive as you think.

Logos acquire meaning through repeated association. A mermaid has nothing to do with coffee, except at Starbucks. This means that your logo is most likely to be effective at a small regional meeting, where people know your institution. At a big national meeting with thousands of posters (where people are most likely to insist on "promoting their

FIGURE 16.4

Visual hierarchy in title bars.

TITLE BAR WITH **WEAK** VISUAL HIERACHY

 Arachidonic acid anomalously accumulates after archetypic apoptosis at aardvark association areas

Anna Author, Aaron Associate, and Alana Advisor
Arizona Affliation

TITLE BAR WITH **STRONG** VISUAL HIERACHY

Arachidonic acid anomalously accumulates after archetypic apoptosis at aardvark association areas

Anna Author, Aaron Associate, and Alana Advisor. Arizona Affliation brand," because they don't understand the difference between a brand and a logo), it's most likely pointless.

Increasingly, projects are collaborations between many participating institutions. If each demands its own logo on the poster, the clutter factor increases a lot. The likelihood that the logos will have complementary colors or shapes is almost zero.

If you want to include logos, do them right. I most commonly see two (or more) of the same logo bookending the title, and those poor logos are squashed, misaligned, pixelated – and drawing attention to all these faults thanks to the white box around them.

It's understandable how this happens. People center titles because they think centering looks classy. This leaves space on either side, and people think that every blank space should be filled. They look for a logo, grab the first thing they can find off their university's web page, and stick it in the corners. The logo is the wrong shape, so people resize it without paying attention to the original proportions. It's understandable, but it's still sloppy.

Make sure your logos, like all your images, are high resolution. Do not grab images that are only a hundred pixels wide from your institution's home page on the internet and blow them to several inches across. Institutions usually have high-resolution versions of their logos and style guides for their use. Sometimes those logos will be on the university's website, but sometimes they will be someplace less obvious, like in a public relations office or with a graphic design team, and you will have to ask for them. Many institutional logos come in alternative colors (full color, limited colors, or grayscale) and shapes (designed to fit in vertical and horizontal spaces), which can help harmonize the logo with your poster, including with other logos.

If you have a colored background on your poster, take the time to make the logo transparent (see *Image types* in Chapter 8). This will often mean converting from JPG to PNG. Better still if you can find a vector-based image.

Don't repeat logos. The redundancy (affiliation shown twice with the logos, once with the text) is pointless. In some cases, using a logo twice, on either end of the title, forces a one-line title to split over two lines, taking up valuable space where the results and data could be.

If someone insists you use the logo, there are alternatives to bookending the title. You may find a home for the logo at the bottom, the "fine print" zone in the lower right corner (Duarte 2009a, Purrington 2019). Here, logos can serve to even out some unused spaces on the bottom of a poster. Another option is to use standard left alignment for the title to create white space to the right where the logo can sit far from the text and be less distracting.

As someone who likes graphics, I love logos and appreciate the skill that goes into making them. But I have seen them used so badly so often that the best option for logos is usually to use none.

Selfies

Occasionally, presenters put headshot-style portrait photographs of themselves on their posters. In theory, having your picture on your poster makes it easier for your audience to recognize and find you, particularly if the poster is hanging long before the actual poster presentation time. Plus, people are naturally drawn to looking at faces.

In practice, headshots are tricky. If you want someone to recognize your face, the photo should be large and high, probably in the title bar. As with logos, it can be difficult to justify devoting some of your poster's precious and limited space to a photograph. If the poster is not hanging up before the presentation time, and you are standing by your poster the entire time, it's unlikely that anyone will be confused about who is presenting the poster.

Because many posters are coauthored, questions of perceived credit can rear up again. While you no doubt discussed bylines when submitting the title and abstract (see *Discuss with colleagues* in Chapter 5), some collaborators might feel snubbed if only one person's picture is on the poster. Pictures of large groups of people make the faces so small that they won't help a viewer identify anyone, much less the presenter.

Putting your own photo on a poster can look somewhere between a little too eager and downright vain. It suggests that you are more interesting or important than the work. Ultimately, people are more interested in seeing themselves in your work than they are in seeing you (see *The conversation* in Chapter 24).

If you do want to include your own photo, it is probably a good idea to have someone else take it; perhaps consider hiring a photographer (see Chapter 5). Many professional photographers specialize in portraits, do them all the time, and understand how to make faces photogenic.

Taglines

In magazines, titles are often followed by taglines – maybe one or two sentences to give a reader a feeling of what the article is about. "A good tagline can help reassure site visitors that they are in the right place, and provides context for the detailed content" (Loranger and Nielsen 2017).

In *Science* magazine, an article titled "Troubled treasure" (Sokol 2009) ran a tagline underneath it: "Mined in a conflict zone and sold for profit, fossils in Burmese amber offer an exquisite view of the Cretaceous – and an ethical quandary." The author's byline comes underneath that. In *Starlog* magazine, the article "Buck Rogers becomes the movie" (Houston 1979) had a tagline, "Covering an SF production from its inception is a fascinating experience, much like watching the growth of a living thing. Sometimes its evolution is predictable, but sometimes the final shape of a show surprises even those responsible for giving it life."

These are rarely used on posters, but they can be an effective alternative to an abstract. A tagline should follow the title immediately, above the authors and affiliations.

Chapter recap

- The title is the most important part of the poster, because it is all that most people will ever read.

- People care a lot about poster titles, but very little about other information that is usually stuck in title bars.

- Your institutional logo is less interesting than you think it is.

CHAPTER 17

Blocks of text

Remember the design brief (Chapter 4): you are competing for attention and people are busy.

You only have a single canvas for a poster. Everything must fit within that space. In contrast, it's rare to have only one slide. Slides are gregarious things, preferring to appear in swarms. With digital slides now the norm, you effectively have infinite space (even though you see only one small section at a time). A poster will always have more elements to consider than any one slide. This presents a bigger challenge to layout and makes greater planning demands.

Slides always accompany a speaker, and explicitly serve as aids to a presentation. You never walk into a room at a conference where there's just a series of projectors, with slides flicking by on auto advance. But a poster needs to be understandable when the presenter isn't there to explain it. One of the biggest struggles in designing a poster is to decide how much can be removed without making it impossible for someone to understand the poster when the presenter isn't around.

Writing the poster

Deciding what you want to include on a poster is always hard, but it's often particularly tough when it's a "*first.*" It might be the first time a presenter is going to an academic conference, or the first time this project is being presented publicly. If it's the first time you're committing ideas and data to paper, you tend to think it through by writing it out. It's no accident that the word "essay" comes from French for "try" or "attempt" – you're trying to clarify your thinking by writing it out. The result is often a poster with lots of words (FIGURE 17.1). This thoroughness might be useful when you are ready to write a journal article about your project, but a poster will be the worse for it. Do not think of your poster as a first draft of a journal manuscript. The two are very different forms of communication. If your poster is your first presentation of a project, you must be **ruthless** in editing it.

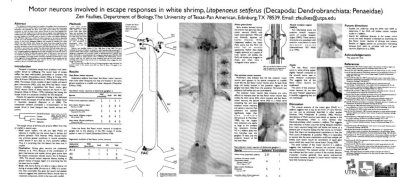

FIGURE 17.1

A poster with too much text.

That conference posters are intended for peers lets you use a few shortcuts in communication (Duarte 2019b):

> **Use familiar language.** To get your own team or peers on board, you need to speak your geek. You probably already have common goals and a common language … It's okay to use the visual and verbal shorthand your team uses on a day-to-day basis. Acronyms, departmental verbiage, and complex charts are all okay, just as long as they are familiar to all involved.

How much you can remove depends a lot on the poster-session scheduling (Gurwell 2012). For some conferences, the posters go up and stay up throughout the entire conference. There is no way a presenter can be there all the time, and audience members will stroll in during a coffee break, at lunch time, or when they need to chill out. Then, a poster should be self-contained, and you may need to write out more to answer common questions.

At other conferences, there is a single viewing block where you are potentially standing by your poster the entire time. The poster can focus on the main question and results, and you can answer the common questions that people ask.

Message discipline

Sometimes, people say "less is more" with posters, but I prefer saying a poster must have "message discipline." Good posters are disciplined. Good posters are focused. They get to the point without dawdling, meandering, or waffling.

Because posters have limited space and are competing for attention, the text on your poster should be concise. The poster should be a focused document with as few messages as possible. A poster is not a place to put tangents, fiddly bits, and details. Save those for the conversations you have with viewers when you present your poster. The phrase "message discipline" is useful because it emphasizes that it is a difficult thing to do. We all want to talk about more things than we have time for. It takes practice to figure out what your message is and stick to it. Thinking about narrative and using ABT structures (see Chapter 6) helps, but some of the message discipline comes during writing and editing. Cutting even one little word can make a paragraph a whole line shorter. And shaving a single line from a paragraph can make a significant difference in making everything fit!

Some people will object that this is a push to "dumb down" a poster, but it is not. Simple does not mean stupid. Most writing problems can be solved by making things simpler. Guides with advice on being concise are worth reading, regardless of publication date (Flesch 1946, Orwell 1946, Strunk and White 2000).

It might be shocking to be told to write simply. In grade school, you might have been regularly encouraged and rewarded for using big or rare words. This is understandable, because it's important to expand your vocabulary at that age. But the goals are different in technical writing. People who use complex words are considered less intelligent than people who present the same content using simple words (Oppenheimer 2006). Technical writing is complicated enough because the topics we write about are complex. It doesn't need to be made more complicated by the writing style.

George Orwell (1946) wrote, "Never use a long word where a short one will do." Luckily, we usually know short words, even if we've been conditioned to use long ones, either by explicit training or by reading too many technical papers. TABLE 17.1 shows some examples.

Orwell continued his advice with, "If it is possible to cut a word out, always cut it out" (Orwell 1946). One of the most common reasons for adding in extra words is stock phrases. We say lots of phrases as "filler" to avoid awkward pauses while talking, and the habit continues when

TABLE 17.1

Long words and their shorter replacements.

Long word	Short word
demonstrate	show
utilize	use
novel	new
sufficient	enough
extant	exist
acquire	get
facilitate	help
quantify	measure

we're writing. Many writers want to record an idea before it flits away and escapes (Gopen and Swan 1990). Stock phrases let you keep typing while you're deciding how a sentence is going to end, letting the physical act of typing flow uninterrupted. The stock phrases just seem come out of your fingers when you type, just as we use them in speech. But after the idea is down on the page, the stock phrases have done their job, add nothing to the text, and can be cut.

It takes practice to recognize and cut stock phrases out of your writing. After revising your own text often enough, you start to recognize your own stock phrases. I'm at least aware that I use "the fact that" too much because I've had to cut the word out of my writing so often.

TABLE 17.2 contains a sample list of stock phrases that can be replaced with shorter words or phrases. Once you become attuned to stock phrases ("the fact that," "is known to be"), it becomes easier to recognize them, cut them out, and replace them without losing any meaning ("That," "is").

Many of the phrases revolve around the choice of verbs. Watch for variants of "to be." Academics often use "known to be" or "shown to be" instead of "is."

Sentences that begin with "What" or "There" can often be shortened. "There are many ways to store data electronically" (eight words) can be revised to "Data can be stored electronically many ways" (seven words). "What is important is this gene" can be "This gene is important." "In many data sets, there are …" can be "Many data sets have …"

Another way to shorten text is to look for nouns that can be replaced by verbs. For example, "caused a reaction" can be replaced with "reacted." More are shown in TABLE 17.3 (Lang 2017).

Another tactic for shortening text is to prune adjectives and adverbs. "Really," "very," and many words ending in "–ly" can often be removed without changing the meaning of the sentence. The prime place to look for these modifiers is in the introduction and results sections, because that is where you are most likely to be add in qualifiers like "probably" or "possibly" to look like a cautious and nuanced academic.

Sometimes, length comes from the overall sentence structure rather than particular words or short phrases. The typical sentence structure is "subject, verb, object." We expect to see the subject at the start of the sentence, with the verb following the subject. As with the elements of a poster, people have to put in more effort to understand a sentence when its parts are in unexpected places (Gopen and Swan 1990). In this example (Gregory 1992), the main topic is buried halfway through the middle of the sentence:

TABLE 17.2

Phrases and
their shorter
replacements.

Long phrases	Short words
in connection with; concerning the matter of; with reference to; in regards to	about
pertaining to	about, of, on
relative to	about, on
in relation to	about, to, with
successfully accomplish	accomplish
afford an opportunity	allow, let
in addition	also, besides, too
in spite of	although, despite
at all times	always
in the amount of	amounting to, for
is applicable to	applies to
on the grounds that; as a result of; due to the fact that	because
in advance of; prior to	before
by means of	by
is authorized to	can
for the duration of	during
adequate number of; sufficient number of	enough
with the exception of	except
comply with	follow
for a period of	for
provide(s) guidance for/to	guide(s)
is responsible for	handles
has an adverse effect on	hurts
in the event that; in the event of; assuming that	if
at a later date	later
is known to be	is
similar to	like
a large proportion of; a large number of	many
a greater degree of; a greater proportion of	more
a majority of	most
it is essential that	must
in close proximity	near
has a requirement for	needs
at this point in time; at the present time; at present	now
in a timely manner	promptly
in the near future	soon
in order that	so
the fact that	that
in order to; in an effort to; as a means to; for the purpose of	to
until such time as	until
make use of	use
at the time that	when
mouse model	mice
in the process of	(Omit)
take action to	(Omit)
the month (or year) of	(Omit)
the use of	(Omit)
in other words	(Omit and remove what preceded it)

Noun	Verb
reaction	react
passage	pass
removal	remove
assistance	assist
employee	employ
dependence	depend
writing	write
organization	organize
assignment	assign
failure	fail

TABLE 17.3

Nouns that can be replaced by verbs.

> No avicide myself and, indeed, not much of a wide-ranging aviphage, I had always assumed that the so-called glorious twelfth occurred only in August when the aristocratic victim of your matched pair of Churchills or Boss' or Purdies (or what have you at £12,000 a throw) is, of course, our only and uniquely indigenous bird, the red grouse.

The topic of that sentence is the "glorious twelfth." The rest of the sentence is distraction from learning that "glorious twelfth" happens in August. But at least the verb "occurred" immediately follows its subject. Not so in the example below (Gopen and Swan 1990):

> The smallest of the URF's (**URFA6L**), a 207-nucleotide (nt) reading frame overlapping out of phase the NH_2-terminal portion of the adenosinetriphosphatase (ATPase) subunit 6 gene **has been identified** as the animal equivalent of the recently discovered yeast H^+-ATPase subunit 8 gene.

The subject ("URFA6L") and verb ("has been identified") are separated by more than half the sentence.

We are familiar with the "subject, verb, object" structure, which explains why sentences written in passive voice are frustrating. The passive voice creates a sentence structure that is in effect "object, verb, subject." Sometimes, the sentence is "object, verb" and the subject is only implied. This usually ends up being longer because the sentence usually needs some version of the verb "to be" to make sense (Lang 2017). If you can add the phrase "by Daleks" to a sentence and it still makes grammatical sense, the sentence is written in the passive voice and should probably be revised. For example, "The poster was hung on its specified board" still scans when it becomes "The poster was hung on its specified board by Daleks." Using the active voice – "I hung the poster on its specified board" – makes it easier to read. "The passive voice dissolves the power of narrative. It destroys the impact of action. It sows confusion about who did what" (Zimmer 2019).

That said, the passive voice can be appropriate, as in this example from Gopen and Swan (1990):

> "Bees disperse pollen" and "Pollen is dispersed by bees" are two different but equally respectable sentences about the same facts. The first tells us something about bees; the second tells us something about pollen. The passivity of the second sentence does not by itself impair its quality; in fact, "Pollen is dispersed by bees" is the superior sentence if it appears in a paragraph that intends to tell us a continuing story about pollen.

Look for lists that can be combined. For example, consider this example as a summary statement: "In drought years, there was no change in nest survival or fledging success, but nest orientation shifted south, nest initiation peaked early, and nest predators were more diverse." Both "nest survival" and "fledging success" measure reproduction. Both "orientation" and "initiation" describe nesting behavior. Both examples can be combined into a single category. Here is a suggested revision: "During drought, birds maintained reproductive success despite changed nesting behavior."

Editing your own work is hard, but it gets easier with practice. I still catch myself typing "the fact that" and "really" all the time, despite years of trying to remove them from my writing. But at least I know enough to look for them now.

Bullets

People who are trying to be concise sometimes try to compress all their relevant text into bullet points, and "use bullet points" is a common recommendation for poster makers. But there are more negatives to using bullets that positives.

Bullet lists destroy narrative. Bullets are impoverished in showing relationships compared to sentences. "Lists can communicate three logical relationships: sequence (first to last in time); priority (least to most important or vice versa); or simple membership in a set (these items relate to one another in some way, but the nature of that relationship remains unstated). And a list can show only one of those relationships at a time" (Tufte 2006). Sentences arranged in paragraphs is the style we read most often, and it's how people expect complicated ideas to be presented.

Formatting inconsistency is another common problem with bulleted lists. First, software often handles the size and spacing of bullet points poorly,

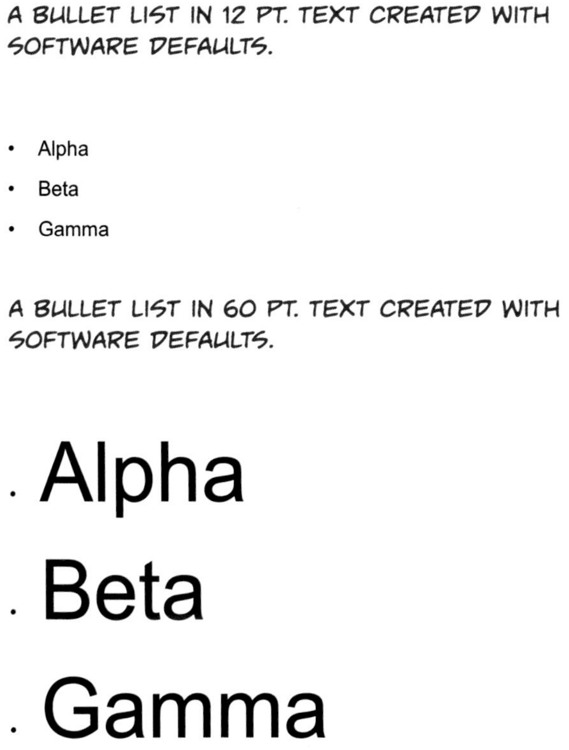

A BULLET LIST IN 12 PT. TEXT CREATED WITH
SOFTWARE DEFAULTS.

FIGURE 17.2

Bullet lists that
scale poorly
with text size.

- Alpha
- Beta
- Gamma

A BULLET LIST IN 60 PT. TEXT CREATED WITH
SOFTWARE DEFAULTS.

. # Alpha

. # Beta

. # Gamma

particularly when you move to large point sizes (FIGURE 17.2). Second, people are inconsistent in style. All too often, a list is a mishmash of complete sentences, phrases, and single words. Third, people are inconsistent in how they type bullet lists. For example, people sometimes punctuate some bullet points with a period, but leave others points in the same list without a period. When people write sentences in paragraphs, they are much better about putting a period at the end of every sentence.

This is not to say that bulleted lists are useless. They are appropriate for short lists. A poster, though, should be more than just short lists.

Pull quotes

Pull quotes are frequently used in newspapers and magazine. They are short excerpts "pulled" from a larger piece of text, one or two juicy bits in a story set off from the rest of the text in large type to give a bit of a teaser as to the content of the article or interview. But apart from the occasional descriptive section heading, I have never seen them used in posters.

Some posters don't lend themselves to images, but you might be able to turn plain text into something more graphic. Select a particularly attention-grabbing phrase or sentence. Then, set it in very large type to make it stand out from the rest of the text. Make it so much bigger than the main text that people will see and read the pull quote first, previewing the rest of the text and encouraging people to read on.

It is best if the pull quote is placed before where it appears in the full text, however. For example, if the pull quote appears in the second column, the actual quote text it is quoting should not be in the first column. It feels anticlimactic to be reading, come across the pull quote, and think, "I read that already."

Pull quotes can be placed like graphic elements within text (FIGURE 17.3). They can be placed in line with the main copy, or between columns, with text wrapping around them.

The fewer the characters, the more "picture-like" text becomes. With a little work with typography, text can become a graphic element (FIGURE 17.4). Such playfulness and experimentation with type can be a way to add visual interest to posters that are driven by textual analyses rather than numeric data.

FIGURE 17.3

Pull quote.

A PULL QUOTE LETS READERS KNOW THERE'S JUICY STUFF IN THESE GREY BLOCKS OF TEXT!

Even as I beheld this a lurid green glare lit the road about me and showed the distant woods towards Addlestone. I felt a tug at the reins. I saw that the driving clouds had been pierced as it were by a thread of green fire, suddenly lighting their confusion and falling into the field to my left. It was the third falling star!

Close on its apparition, and blindingly violet by contrast, danced out the first lightning of the gathering storm, and the thunder burst like a rocket overhead. The horse took the bit between his teeth and bolted.

A moderate incline runs towards the foot of Maybury Hill, and down this we clattered. Once the lightning had begun, it went on in as rapid a succession of flashes as I have ever seen. The thunderclaps, treading one on the heels of another and with a strange crackling accompaniment, sounded more like the working of a gigantic electric machine than the usual detonating reverberations. The flickering light was blinding and confusing, and a thin hail smote gustily at my face as I drove down the slope.

"And this Thing I saw! How can I describe it? A monstrous tripod, higher than many houses..."

At first I regarded little but the road before me, and then abruptly my attention was arrested by something that was moving rapidly down the opposite slope of Maybury Hill. At first I took it for the wet roof of a house, but one flash following another showed it to be in swift rolling movement. It was an elusive vision—a moment of bewildering darkness, and then, in a flash like daylight, the red masses of the Orphanage near the crest of the hill, the green tops of the pine trees, and this problematical object came out clear and sharp and bright.

And this Thing I saw! How can I describe it? A monstrous tripod, higher than many houses, striding over the young pine trees, and smashing them aside in its career; a walking engine of glittering metal, striding now across the heather; articulate ropes of steel dangling from it, and the clattering tumult of its passage mingling with the riot of the thunder. A flash, and it came out vividly, heeling over one way with two feet in the air, to vanish and reappear almost instantly as it seemed, with the next flash, a hundred yards nearer. Can you imagine a milking stool tilted and bowled violently

196   BETTER POSTERS

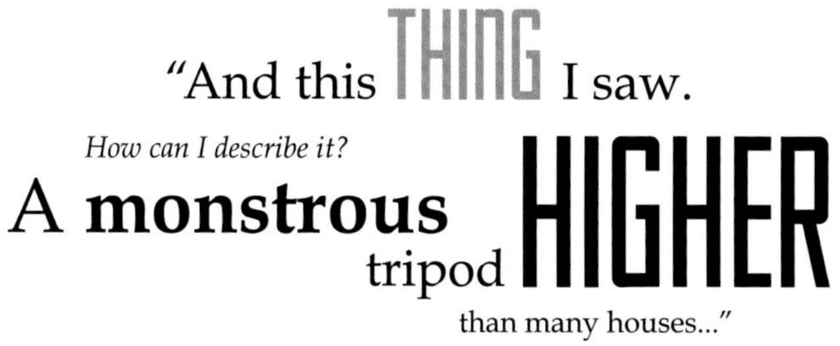

"And this THING I saw. *How can I describe it?* A **monstrous** tripod HIGHER than many houses..."

FIGURE 17.4

Graphic typography.

Abbreviations

It is tempting to use abbreviations on a poster. After all, the point of an abbreviation is to save space. But the amount of space saved is not worth the confusion that abbreviations cause in the reader. You as an author might have weeks, months, or much longer to become learn an abbreviation, but a poster viewer may be encountering it for the first time. The viewer must hold the meaning of the abbreviation in short-term memory, or continually look somewhere else on the poster to be reminded of what it means.

Worse, no matter how sensible the abbreviation, there is always the possibility that the viewer is more familiar with another meaning for the same string of letters. For example, "raised-visual signs" might be "RV," but many readers might already know "RV" as "recreational vehicle." You may want to refer to a stretch receptor as "SR," but the reader might think of "stimulus and response." Depending on who you talk to, an "RPG" could be a "role-playing game" or a "rocket-propelled grenade," which are two things that should never be confused.

A few abbreviations are so widely known and used in common culture that it is probably more confusing to spell out the term. Thanks to crime shows, home ancestry testing, and popular media, more people know "DNA" as genetic material than "deoxyribonucleic acid."

Point size

A common question from novice designers is, "What is the minimum point size for poster text?" This is going about things the wrong way. It's working from, "How much can I put on the poster?" Flip this. If you have made careful decisions about what you want to include, and focused your poster, you should be asking how big you can make your text, not how small your text can be.

On a more practical level, "point size" gives you only a very crude idea of how much space a block of text will take up on a page. And that point

FIGURE 17.5

Point size and
text legibility.

THESE ARE THE SAME POINT SIZE, BUT THEY DO
NOT HAVE THE SAME MINIMUM *READABLE* POINT
SIZE!

The cross between the Sebright and the Black-Breasted Game bantam was undertaken primarily to study the inheritance of hen-feathering. The Sebright was chosen, on the one hand, because this race is pure for hen-feathering, whereas in other races, such as the Campines, both kinds of males are known. The hen-

The cross between the Sebright and the Black-Breasted Game bantam was undertaken primarily to study the inheritance of hen-feathering. The Sebright was chosen, on the one hand, because this race is pure for hen-feathering, whereas in other races, such as the Campines, both kinds of males are known. The hen-feathered birds of such races

size does not say anything about how type will look on the page or how readable it will be (FIGURE 17.5).

When making decisions about whether the point size is big enough, there is no substitute for printing sample of text at full sized, sticking it on a wall, stepping back, and seeing. Then putting your print sample in low light and looking again. Then bring in someone who might not have eyesight as good as yours and asking them to look at it.

But if you must have a number for minimum point size, many parks and museums use text with a 4.5 mm x-height (about 24 point) as the

FIGURE 17.6

Minimum text sizes
used by museums
and parks.

EXHIBIT STANDARDS FOR PARKS AND
MUSEUMS

MINIMUM SIZE FOR *ANY* VIEWING DISTANCE:

Wax   4.5 MM X-HEIGHT (~24 POINT)

VIEWING DISTANCE OF 6 FEET (2 METERS):

Box   19 MM X-HEIGHT (~100 POINT)

VIEWING DISTANCE OF 10 FEET (3 METERS):

Fix   28 MM X-HEIGHT (~148 POINT)

minimum point size for exhibit labels, including those that can be read mere inches away (FIGURE 17.6). For exhibits viewed from 6½ feet (2 m) away (equivalent to typical poster viewing distance), parks and museums use text with a 19 mm x-height minimum (about 100 point) (Smithsonian Accessibility Program n.d., Parks Canada 1993, Harpers Ferry Center Accessibility Committee 2017).

The Americans with Disabilities Act also has standards for character size that can be helpful as a guide for posters (Department of Justice 2010; FIGURE 17.7). Those standards measure text by cap size, specifically capital I. For a sign meant to be viewed from 6 feet (1.8 m) or closer, the cap height of text should be at least 16 mm (⅝ inch; about 66 points, depending on the font). For each additional foot (0.3 m) of viewing distance, the letter size should increase by 3.2 mm (⅛ inch). For a viewing distance of 10 feet (3 m, chosen because it's a round number), text should be 28.8 mm (1⅛ inch, or about 120 points).

While these standards provide reasonable starting points, it is not always clear how the recommendations are derived. Some standards are based more on intuition than on data (Arditi 2017). But these standards have been used for decades in multiple countries, which indicates that they work in practice.

Because these are standards designed to maximize accessibility, these text sizes are much larger than minimum sizes sometimes recommended for posters (Alley 2018).

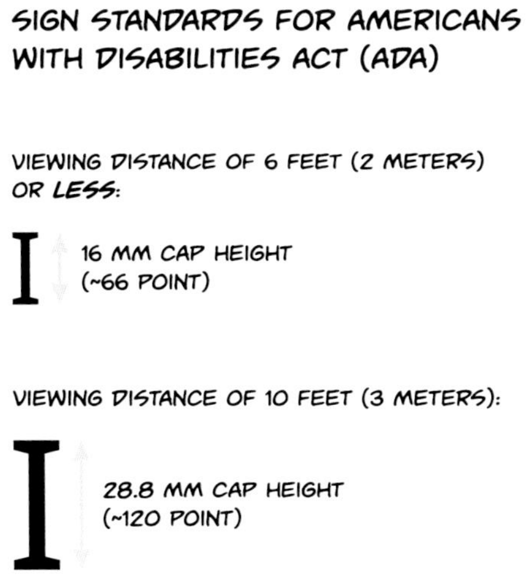

SIGN STANDARDS FOR AMERICANS WITH DISABILITIES ACT (ADA)

VIEWING DISTANCE OF 6 FEET (2 METERS) OR **LESS**:

I 16 MM CAP HEIGHT (~66 POINT)

VIEWING DISTANCE OF 10 FEET (3 METERS):

I 28.8 MM CAP HEIGHT (~120 POINT)

FIGURE 17.7

Minimum text sizes for signs specified by the Americans with Disabilities Act.

Paragraph alignment

Most blocks of text longer than a sentence are set so that the left edge of the text forms a line (ragged right) or so that both the left and right edges of the text form a line (justified). Either one of these can work. Full justification makes it very clear when you've done a great job of alignment, enhancing the sense of order you get with a grid. But full justification sometimes creates extra-wide spaces between words or letters that makes for difficult reading. And the narrower a column, the more irregular the spacing becomes (see FIGURE 14.4). Because of this, ragged right is often recommended for accessibility reasons (Parks Canada 1993). The gaps created by full justification can often be fixed if you're willing to put in the work of fine-tuning the text by hyphenating it. More on this later.

In some software, text can be wrapped around objects. If you do this, have the object on the right side of the text, so that when someone is reading, they can more easily find the start of the next line (Parks Canada 1993).

Main blocks of text should not be centered or right-justified.

Chapter recap

- It is easy and tempting to write too much. Edit your poster ruthlessly.

- Bullet points are concise but destroy narrative.

- Abbreviations should be limited to terms that are either in common use (e.g., DNA) or are extremely common in the field.

- Bigger text is better for everyone.

CHAPTER 18

Sections

Many academic journals have introduction, methods, results, and discussion sections (sometimes called the IMRAD format). It's a strong narrative structure, but it's the default for most journals. As discussed before, falling back on defaults can mean missed design opportunities.

An alternative to the IMRAD format is the inverted pyramid (Visocky O'Grady and Visocky O'Grady 2008). This is common in new reporting. Items are ordered from the most important (the lead or lede), followed by supporting information, with details last. This means that if someone stops in the middle, they are not left with unanswered questions. This format helps ensure people get something valuable even if they don't read all the way to the end.

Section headings

If you use headings, remember to create a strong visual hierarchy. Make the headings bigger than the main text and set them in a different typeface from the main text. Keep the headings close to the main text they relate to, and far from unrelated text.

One advantage of having a very common default format is that people can be so used to it that you don't need to follow the format strictly. Anyone with experience of academic papers will know in general what is supposed to be in each section, so they can "fill in the blanks" of the shape (FIGURE 18.1). Academics are so used to the IMRAD structure that you can change the headings significantly without confusing readers.

THERE IS NO SQUARE HERE. PEOPLE FILL IN GAPS USING THEIR EXPECTATIONS.

YOU CAN LEAVE OUT THINGS LIKE HEADINGS WITHOUT CONFUSING PEOPLE.

FIGURE 18.1

Implied square.

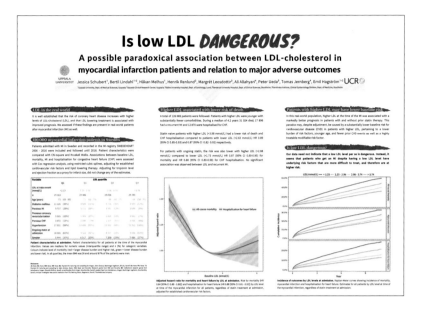

PEOPLE ARE USED TO THIS...

Introduction

We have a question. Lorem ipsum dolor sit amet, consectetur
adipiscing elit. Morbi rhoncus vestibulum scelerisque. Morbi laoreet
vestibulum neque. In feugiat aliquam velit. Pellentesque vitae tristique
leo.

Methods

Here's how we tried to answer it. Nulla eu diam ut magna viverra
facilisis. Suspendisse sed pulvinar quam. Nunc quis dignissim risus, nec
ullamcorper libero. Proin ultrices vitae magna vitae viverra.

BUT THIS...

This is the question.

Lorem ipsum dolor sit amet, consectetur adipiscing elit. Morbi
rhoncus vestibulum scelerisque. Morbi laoreet vestibulum neque. In
feugiat aliquam velit. Pellentesque vitae tristique leo.

We did this experiment.

Nulla eu diam ut magna viverra facilisis. Suspendisse sed pulvinar
quam. Nunc quis dignissim risus, nec ullamcorper libero. Proin ultrices

OR THIS WORKS FINE, TOO!

We have a question. Lorem ipsum dolor sit amet, consectetur
adipiscing elit. Morbi rhoncus vestibulum scelerisque. Morbi
laoreet vestibulum neque. In feugiat aliquam velit. Pellentesque vitae

Here's how we tried to answer it. Nulla eu diam ut magna viverra
facilisis. Suspendisse sed pulvinar quam. Nunc quis dignissim risus,
nec ullamcorper libero. Proin ultrices vitae magna vitae viverra.

You might consider headings that are more descriptive than the stock phrases used by journals, such as in FIGURE 18.2. Instead of an "Introduction" heading, you might write the hypothesis you are testing. Instead of a "Methods" heading, you might write the name of the method you used.

Alternatively, you might drop headings entirely (FIGURE 18.3). You might approach the poster text like an abstract, where the content of those sections is present, but not broken apart by headings. People will understand the structure without the headings being explicitly written on the poster. You could signal the sections of the paper using other cues besides writing out words. You could break apart sections with large capital letters (as in FIGURE 7.6), which increases reading in advertisements (Ogilvy 1963).

If the text is short, you might leave out headings and just use white space to separate sections.

Abstract

Don't use an abstract on the poster. Putting an abstract on a conference poster is like writing a haiku about a limerick.

Abstracts serve no purpose on a poster. Abstracts were created for situations where you could not access a complete paper. In the distant pre-internet past, academic articles were only printed on paper in journals and sent to libraries. The only way to search for articles was to use abstracts that were collected in big volumes on paper (like *Psychological Abstracts* or *Zoological Record*) so that you could decide whether you needed to traipse through the library to get the whole paper. As journals moved online, abstracts became the preview for subscription journals, helping people to judge whether they wanted to get a paper through interlibrary loan or pay a fee to download the complete paper.

On a poster, the full work is contained on the poster. There is no reason to summarize what someone can read in full by turning their head slightly.

Instructing poster presenters to have an abstract seems to arise out of an unrelenting desire to make everything follow the format of academic papers. But posters are not manuscripts. Conference posters should be self-contained summaries of work. Asking for an abstract does a disservice to presenters, who are forced to stick up more words on a format that works best with very few words.

There are a few possible cases where an abstract might be helpful.

For example, there is typically a single official language at a conference, and it is usually English. However, if the conference is in a location where the main language is not English, and many attendees speak the local language, there may be a good reason to create an abstract in that language.

If you are going to be in a poster competition and know that having an abstract is one of the judging criteria, an abstract is very helpful to you and the judges, if not to the typical conference attendee (Chapter 5).

Introduction

Visitors to your poster are interested and want to know what question you are trying to answer, but they are busy and distracted, therefore you should be able to tell them what your poster is about in a sentence.

The structure of that sentence is one we have seen before (in Chapter 6), and is a reminder that an "and, but, therefore" (ABT) structure can be a very effective way to structure an introduction. Many academics fall back on "and, and, and" (AAA) sentences when trying to summarize their poster, but a series of facts is not compelling.

Methods

Most of the time you do not need to specify the level of detail that someone would need to replicate a technique. If someone wants the full details, they can email you. The exception is if the point of the poster is to present a new method. But if you are using standard methods, this section of the poster can be extremely short. You do not have to show everything you did. You can often put a lot of methodological details in captions of graphs in your results.

Some people like using flowcharts to show methods (see Chapter 9), which is very effective if you can do it. There are fewer software packages that streamline making flowcharts than other technical graphs. If you don't have one of those, making flowcharts can be complex.

Results

You do not have to show everything you discovered. The temptation to show off a lot of hard work is strong, but you need to be willing to edit.

Likewise, your results do not have to be equalized. If you have four graphs, not all four may be equally important. One might show the main effect, while the other three are necessary controls to rule out alternative explanations. If you have a spectacular result, make it the centerpiece. Make it big. Put it first. Put it at eye level. Make the other graphs smaller. They can go further down below eye level.

If your results mostly consist of graphs, put a statement above each graph saying what it shows. Don't bury the interpretation in a legend underneath the graph.

Take-home message

Many posters, because they are so heavily influenced by the standard scientific IMRAD format, end with a discussion section. Discussion sections can include a lot of different things, including why this study advances our knowledge, limitations of the study, suggestions for future directions, and much more. Sometimes, posters end with paragraphs of text. That's too much stuff.

In Chapter 5, I suggested that your title can act as the take-home message for the whole poster. If you didn't structure your title that way, this is the place to say what is the single most important thing on this poster. Ask what one thing you want people to remember after they've walked away from your poster. And remember that it is one take-home message (singular), not many take-home messages (plural).

References

People disagree on whether references are necessary for posters. Some people argue that a poster without references is an automatic failure, because references are one of the defining features of academic writing. Others argue that references are not needed, because a poster is a relaxed format and space is limited. Most people fall somewhere in the middle and show a few key references, because that indicates that you've done due diligence in reading the literature.

If you do include references, you do not need to include as many as a full manuscript would have. A poster is not a complete literature review, so be selective. Audiences will rarely want to copy down the references to check up on them, which is often a major point of references in a paper.

Besides cutting down on the number of references, there are many ways to make each individual citation more compact. Many academics are used to citations in the main body of the text, and a list of references at the end. There are options to shorten both.

In the main body of the text, you might use numbered citations instead of listing the author(s) and year. Many people who know a research field well despise numbered references because they can recognize many papers from the name of the first author and year alone. In a multi-page article, a reader must flip back and forth to find what article is being cited. This is less of a problem with a poster, because both the citation and reference are on the same page.

Reference lists can be much more concise than the way they often appear in journal articles. There are a couple of ways to identify articles unambiguously. If the references are written out in words, a journal title, volume, and first page number (or article number for online journals) are typically all that's needed to identify a paper unambiguously. You might omit the article title, like some journals do. Instead of spelling out the authors' given names, you might use initials with no period between the letters. You might only show family names and not given names. You might only show the name of the first author and use "*et al.*" instead of listing all the authors. You could omit author names and year entirely. You could even list articles by digital object identifiers (DOIs). The last couple of options are drastic and not recommended. Experienced academics often strongly associate a paper with the name of the first author, so the space saved may not be worth how obscure the reference becomes.

But if you go the route of cutting down references to the absolute bare bones, you might use short references in the main body of the text only instead of compiling a list at the end. For example, "We used dry ice as a stimulus (*PLoS ONE* 7: e39765)" instead of "We used dry ice as a stimulus (Brenner *et al.* 2012)" and then forcing people to hunt around near the end to find the rest of the citation.

References are often considered "fine print" and could be set in a smaller point size than the main text.

Acknowledgments

It's good to recognize people who have helped with your science. People often want to show their gratitude to funding agencies by putting agency logos on the poster. But putting logos in acknowledgment sections causes some of the same problems as when they are put in title bars. Logos can be awkward to fit in the space, and people often carelessly use distorted or low-resolution versions. Acknowledgment sections are at least better places for logos than titles, because acknowledgments are usually at or near the bottom of the poster, where they are less obtrusive and obnoxious.

In addition to people, this section is a good place to list conflict-of-interest disclosures and funding sources. If you are short on space, it is possible that a QR code or some other online resource could provide them (Foster *et al.* 2019).

Like references, acknowledgments can be set in a small point size.

Sidebars

Sidebars show information that is not part of the main thrust of your narrative. They are common in newspapers and magazines, which use them for lists, personal stories, recommendations, quizzes, related events, "how to" guides, and so on. For example, a magazine profile of a restaurant might include one recipe for a signature dish as a sidebar.

Sidebars are uncommon on posters because of space constraints, but they can be used effectively, particularly on larger posters. A statement about whether you care to share your poster on social media (see FIGURE 18.5) is an example of content that is appropriate for a sidebar.

A successful sidebar is self-contained and complements other information on the poster. To avoid confusion, a viewer must understand that a sidebar is not part of the main narrative of your poster, which you can signal in a few ways.

First, give a sidebar its own title. This will help convey that this is not part of one of the standard sections ("introduction, methods, results, discussion"). In FIGURE 18.4, the sidebar on the left – titled "What is the evidence for crustacean nociceptors?" – provides context that is not necessary to understand the experiments, and breaks a long introduction into smaller and more readable chunks. The sidebar on the right – titled "Nociception symposium" – advertises an event at the conference related to the content of the poster.

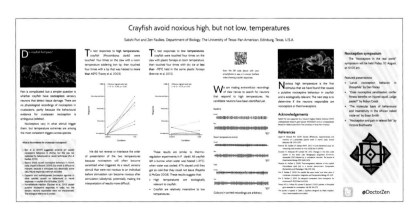

FIGURE 18.4

Example of a poster with sidebars, at lower left and right.

Second, take the name "sidebar" literally: bars on the side. In magazines and newspapers, sidebars can be placed almost anywhere, but those print formats have the advantage of multiple pages and readers that can spend as much time deciphering pages as they want. On a poster, the far right or bottom are probably the best locations for sidebars.

Third, place the contents in a box. This is one of the rare cases where a box is completely appropriate. Heavy black lines tend to be overpowering, so you might try defining the box by filling a rectangle with a color. If the sidebar is large, the color does not need to be very intense to "read" as a distinct color. You can make it just bright enough to distinguish it from the rest of the poster. If you want a bold look, you could try inverting the sidebar's colors. That is, if your poster background is white, your sidebar might be dark with light text.

Social media permissions

Conferences are a weird space. They are public, yet not entirely open. People are expecting to present their work, but it is often preliminary and in an unfinished form, and they may not be ready to share it everywhere, with no reservations, lest it be wrong. So different conferences, and different presenters, have very different attitudes and policies about sharing information presented at conferences. Some conferences encourage it, with official social media hashtags (Shiffman 2012). Other meetings have "no photos" policies, although they are rarely enforced rigidly.

Regardless of where you personally fall on that spectrum, most people at conferences have smartphones with high-resolution cameras, and many are on social media. Because telepathy has not yet been perfected, nobody knows how you feel about sharing your work online except you.

Whether you are comfortable with people sharing your poster on social media or not, don't make people guess. This is particularly useful if your preference is the opposite of the norm for the conference, and for poster sessions where the posters are hung well before people officially present. Put an icon or a notification somewhere on your poster (FIGURE 18.5; Trani 2018).

FIGURE 18.5

Various icons specifying whether viewers can record or share a poster's contents on social media. The right icon includes a Creative Commons license to specify how content can be shared.

Regardless of whether you encourage or discourage sharing of your poster on social media, however, be prepared for the possibility that someone might share it anyway. Do not present work that must remain confidential, for whatever reason.

Chapter recap

- Avoid putting an abstract on a poster unless you're in a competition that requires one.

- The "introduction, methods, results, and discussion" format for academic papers is common but does not have to be explicitly followed by using these headings.

- Include a single "bottom line."

- Be explicit about whether you want people to be able to share your poster.

CHAPTER 19

Images and graphics, revisited

You should by now have picked out what images and graphs you want to show. Now you want to work on refining them for the poster you're making.

Summarize figures

In scientific journals, figure legends traditionally go beneath a figure. For posters, it can be useful to put a summary above the figure, saying what it shows, and put fine details in a more traditional legend underneath the figure if desired.

Harmonize fonts

By now, you should have selected the typeface for the main text of the poster. Make sure the typeface on your graphs is the same as the one your poster is set in. If your poster is set in Helvetica, don't have your axis labels in Times New Roman.

Position and align content

Many software packages treat text boxes differently than other boxes. Many programs add margins around text, but not other items, like photos, lines, and boxes (FIGURE 19.1). This means that while text boxes will align correctly to other text boxes, they won't align with pictures, lines, or any other objects.

In FIGURE 19.2, the images at the bottom left and bottom center are narrower than the column width. This creates awkward white spaces on either side of the image.

If you have several graphs that are going to be near each other, double-check their alignment. One graph might have multiple lines of text for an axis label, while an adjacent graph has a single-line label on the same axis. As shown in FIGURE 19.3, your software might nominally make these graphs the same size, but they look wrong when placed next to each other.

The problem is that the axes are obvious straight lines, while the edges of axis labels are only implied straight lines. The expectation that the

SOME SOFTWARE PADS TEXT WITH MARGINS, SO ALIGNING A TEXT **BOX** DOESN'T ALWAYS ALIGN THE TEXT!

SET TEXT BOX MARGINS TO ZERO OR USE GUIDELINES TO ALIGN TEXT BY EYE!

FIGURE 19.1

Text box padding and alignment with other objects.

FIGURE 19.2

Poster with white spaces around images. (Mary Bratsch-Hines)

FIGURE 19.3

Graphs with
multi-lined axes
labels aligned to
label edges.

IF GRAPHS ARE DIFFERENT SHAPES, DON'T ALIGN LABELS!

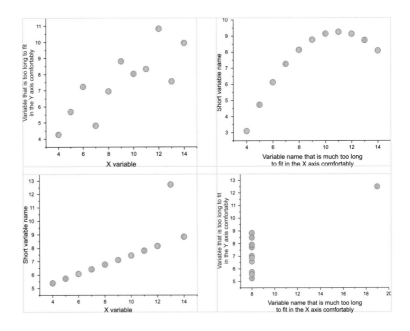

FIGURE 19.4

Graphs with multi-
lined axis labels
aligned to axes.

ALIGN THE *AXES* OF MULTIPLE GRAPHS!

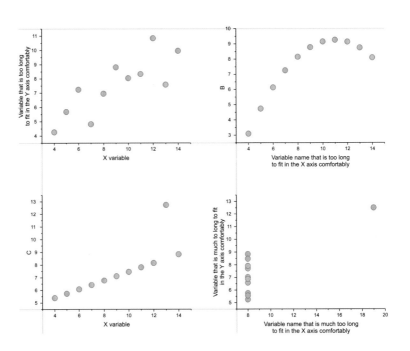

explicit lines will align is stronger than the expectation that the implicit edges will align. As shown in FIGURE 19.4, making the axes line up, even if it means the labels do not, can be much better optically.

For graphs, there is no excuse not to go back to the original file and modify the proportions to fit the space. Photographs, however, are trickier, because you cannot change the proportions the way you can redraw a graph. There are several strategies to make photographs fit into a desired space (FIGURE 19.5). But never, under any circumstances, just stretch the picture.

PROBLEM: THIS PICTURE DOESN'T FIT THE SPACE.

PROBLEM WORSENED: THE PICTURE IS DISTORTED TO FIT.

FIGURE 19.5

Fitting images to available space.

SOLUTION #1: RESIZE AND CROP THE IMAGE TO FIT THE SPACE.

SOLUTION #2: FILL THE SPACE WITH A COLOUR FROM THE IMAGE.

SOLUTION #3: COPY THE IMAGE. BLUR AND LIGHTEN THE COPY. ENLARGE AND CROP THE COPY TO FIT THE SPACE. DROP THE ORIGINAL ON TOP.

Face into the page

If you have any images with people looking somewhere, make sure they look into your poster, not off the edge (FIGURE 19.6). This effect is strongest when dealing with pictures of faces, because we tend to look where other people are looking. But any image that has a strong direction – like an arrow, a pointing finger, or a flying plane – can have the unintended effect of drawing attention away from where you want it.

FIGURE 19.6

Directional cues like gaze influence where people look.

GAZE LEADS OFF THE PAGE

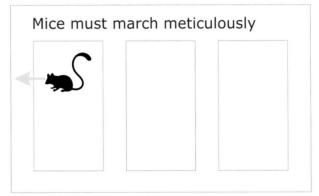

GAZE DIRECTS YOU TO THE REST OF THE CONTENT

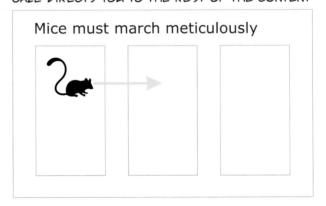

Chapter recap

- Typefaces used in graphs should be the same as in the main body of the poster.

- There are many ways to make pictures fit into a space. Distorting them is always the worst option.

CHAPTER 20

Fine-tuning

You've got a grid, you've got images, you've got text. But just because everything fits in the allotted space does not mean that you're done. Not if you want a poster that is more than just "competent." A great poster is made in the fine-tuning and detail work. As the saying goes, "Trifles make perfection, but perfection is no trifle."

Bottom margins

When laying out text in columns, it's natural to tend to align blocks of text and images using the top left corners. This means that the tops of these elements form a nice, neat line underneath the title. But the amount of content in each column is much harder to judge, and the bottoms of each column may not line up as neatly as the tops of the columns do.

Do not adjust column length by fiddling with size or spacing. People expect those to be consistent throughout the poster. If there is a 36-point gap between a heading and body text in column one, there should not be a 12-point gap between a heading and body text in column three.

The best option for harmonizing column length is to adjust the sizes of graphs and figures. It's not as critical to keep graphs and images the same size throughout a poster as it is for text. Go back to your graphing software to make a graph taller. Open your photo editor to crop a photo to make it narrower or shorter.

If the space left on the bottom of each column is small, just leave the bottoms of the columns a little uneven.

While it seems unlikely that there may be much space left at the foot of a column (given that most posters tend to have too much stuff), large spaces at the bottoms of columns are good places for tangential material. Examples might include your institutional logo, social media permission information, a cartoon, or a sidebar. Even if adding these elements does not make the bottoms align exactly, they can minimize the unevenness.

Hyphenation

If you're using fully justified text, hyphenating can significantly reduce unwanted wide spaces in lines. A quick and dirty general rule is to place hyphens where words break into syllables. But hyphenating is a tricky business, and it's not easy to spell out infallible rules. A four-syllable word like "relationship" could be hyphenated in three different ways ("re·la·tion·ship"). But breaking the word apart into "re-" and "lationship" seems strange compared to "relation-" and "ship." Most professional typesetters don't trust the job to software algorithms and do it by hand instead.

To add even more complexity, many typesetters who are using fully justified text will often set the text so that the hyphens actually sit slightly outside the right-hand edge of the text, preferring to let only letters align (Landa *et al.* 2007).

Symbols

Punctuation marks made from horizontal lines are an example of the need to pay attention to differences in symbols. A hyphen, a minus, and a dash are all distinct symbols (FIGURE 20.1). A hyphen is the shortest and is used to combine or split words. Minus signs are a little longer, have a little more space around them, and are used for equations. Dashes are the longest and are used to set words or phrases apart. Dashes come in two forms. The en-dash is about the width of the letter n and the em-dash is about the width of the letter m. Different people have different preferences about which dash to use, but as long you use one consistently, you should be fine.

The differences between symbols are small enough that English keyboards only provide a combined "hyphen-minus" key, and many fonts don't even

		Sitka	Noto Sans	Cabin	Book Antiqua
FIGURE 20.1	HYPHEN-MINUS	f-f	f-f	f-f	f-f
Hyphens and dashes in different typefaces.	HYPHEN	f-f	f-f	f·f	NO HYPHEN FOR THIS FONT
	EN DASH	f–f	f–f	f–f	f–f
	EM DASH	f—f	f—f	f–f	f — f

WRONG LETTERS	RIGHT SYMBOLS
20°C	20°C
3x2=6	3×2=6
10 um	10 μm

have a minus symbol. But if you are going to hyphenate, use the correct hyphen symbol if there is one.

The differences between hyphens and dashes is a good example of the importance of special symbols. When you're typing out text, you might use ordinary letters to stand in for special symbols. As part of your final checks, go through your text, check for any technical symbols and make sure you have used the proper symbols, as shown in FIGURE 20.2. A capital letter O is not a zero, and a superscript letter O is not a degree sign. A letter X is not a multiplication symbol. A lowercase u is not the metric symbol for micro-. People can tell the difference.

NASA engineers were chastised in the investigation of the *Columbia* Space Shuttle disaster for their inconsistent format of units, as shown in FIGURE 4.2 ("3cu. In," "1920cu in," and "3 cu in"). The board wrote, "While such inconsistencies might seem minor, in highly technical fields like aerospace engineering a misplaced decimal point or mistaken unit of measurement can easily engender inconsistencies and inaccuracies" (Columbia Accident Investigation Board 2003).

Kerning

Now that your text fits, this is the time to make any fine adjustments to the text. Look for any letters that have a lot of diagonals, like A, W, and Y. Fonts vary substantially in how well these letters "fit" together on a page. The spacing between letters may need some adjustment by hand (FIGURE 20.3).

ALMOST NO SPACE

WIDE SPACES

DEFAULT TEXT SPACING
LEAVES VARYING GAPS

WAVE

KERNING BY HAND
EVENS OUT THE GAPS!

WAVE

While some software offers automatic kerning, you will still often have to go in and check and make any adjustments by hand.

You may not have time to kern all the text, but if you have the time, look at the title. The large point size of the title means that any problems are magnified and more obvious there than elsewhere on the poster. Even if there are no unsightly gaps between letters, it can be beneficial to increase the spacing between the letters in titles to make them more distinct and less likely to blur into each other when read from a distance, particularly by people whose eyes have focusing problems.

Take a break

At this point, you have probably been looking at your poster for hours over many days. This is a great time to walk away from the poster for a while.

When you're in the middle of a project that you designed and carried out, everything seems important. But irrelevant details fade away as you spend time away from a project. Think about a favorite album or TV series that you haven't watched in years. You will remember the highlights. You can get clarity by not working on a poster for a few days, then coming back to it with fresh eyes. It will then be easier to spot the embarrassing typos or find ways to improve the poster. This might be a good time to revisit the proofreading tips in Chapter 5.

Writing a script

A poster presentation can be a conversation, with lots of back and forth between the presenter and viewer, but many people will walk up to you and ask you for a "guided tour" through your poster. Once you have a draft poster that includes all the material you want, think about what you're going to say in front of your poster as part of your guided tour. Even better, write a short script for yourself.

Your poster text is different than a presentation script. The poster text is meant to be read. It probably follows the IMRAD structure, with headings and fine print. A script is meant to be performed. It might omit some material so you can hit your main points in a few minutes. And there are often things you want to tell people that don't fit on the poster. The style can be relaxed. You might put in jokes or turns of phrase that wouldn't work in print. You can write prompts to remind yourself to interact with the viewer you're speaking to.

One of the major advantages of writing a script is that you sometimes discover a more logical way to present your narrative than when you first laid it out. For example, you might have a figure that you put on the right side of the poster because it fits in that space. It might make perfect sense when you read through the poster. But when you give people "tours" through the poster at the meeting you might discover that you refer to that picture very early on, in conjunction with something on the left side of the poster. This will force your audience to look over to a different section of the poster, which can be a long way if the poster is big, and other people might be blocking the view. It can disrupt the flow of the presentation.

Writing a script is also helpful to judge how long you will talk. Remember that most people want to spend a maximum of five minutes with you at your poster (see FIGURE 4.8). Your verbal description of what is on the poster should clock in at about four minutes (80% of time) so you and the viewer can exchange pleasantries and ask a question or two. This is not shortchanging yourself. If you respect the viewer's time, they are more likely to feel generous about you and your work and do nice things like recommend your poster or talk to you later if they have follow-up questions. But if it takes twenty minutes for you to read your script, you will wear out your welcome and annoy your audience.

When writing your script, give yourself specific prompts, not just "Explain the graph." Use your script to get into the habit of telling people "which way is up" for every graph (see Chapter 9). For example, "The x-axis shows time and the y-axis shows the amount we measured. We saw a statistically significant increase over time."

User testing

While you are taking a break from the poster, show it to someone else. Show it to an outside viewer who doesn't have the same emotional or intellectual investment in the project that you have. Show it to someone who isn't in your lab. Show it to a non-expert. Show it to someone with a different skill set.

An outside observer is not necessarily, nor should be, an unbiased one. Everyone has their own tastes and preferences and styles. An outside observer may not be objective, but they will at least have different biases than you.

People rehearse oral presentations given with slides. You should do the same for your poster. The moment you hand anything you design to other people, it changes, regardless of how much preparation you make. There is no substitute for criticism and feedback.

Chapter recap

- Small adjustments make a huge difference to the overall impression of the poster.

- Checking and testing a poster is easier if you take a break from working on it. The longer the break, the better.

CHAPTER 21

Before you print

Test legibility

Posters should be readable from about 6½ feet (2 m) away. But how can you test whether it's readable without printing off the entire thing, which might itself be more than 6 feet of expensive paper?

A useful test is to shrink the poster to a letter-sized (or A4) sheet of paper. Most graphics programs will have a "shrink to page" option somewhere in their print menus. Then, hold the printed page at arm's length.

You should be able to read the text and make sense of any pictures or graphs, even at this reduced size.

This won't pick up a lot of problems, like whether an image will be pixelated when printed at full size, but it is a good way to test whether you're trying to cram too much stuff on your poster by making everything too small.

Critiquing posters

Now that you have been through this process, you should able to identify problems in posters. For example, the poster shown in FIGURE 21.1 has multiple places for improvement.

Some of the problems of this design are highlighted in FIGURE 21.2.

1. The title is small. The lines are not evenly spaced.
2. The section headings have no reason to be curved. They also change color every time they appear.
3. There are boxes within boxes, with inconsistent colors and alignment.
4. Random changes in font size in the middle of the sentence.
5. Heading numbering changes from thin to thick.

FIGURE 21.1

Conference poster
for review.

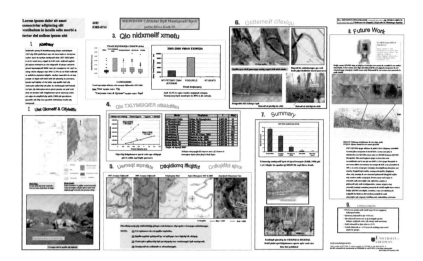

6. Graphs filled with inconsistent patterns instead of solid colors.

7. "Data prison" table has too many vertical and horizontal lines.

8. Wide gaps show lack of alignment between boxes.

9. Numbers in list are underlined using bars instead of true underline.

10. Contact information is small and changes color in mid-sentence.

With experience, you can start to pinpoint areas where things could be done better or at least differently. But there is no end to the number of ways that posters can fail or go right. Early on, you might want to use a checklist to remind you of some of the common pitfalls. The following checklist – inspired by one produced by Nancy Duarte (2009b) – can help you evaluate your poster, and other people's posters, too. It mostly focuses on the graphic elements of the poster, rather than the content. Give yourself a point for every question you answer "yes" to, for a best possible score of fifteen. Have other people do this checklist, too, because you may not be objective after having worked on the poster for so long.

1. Does the poster have an easy **entry point** that lets someone know at a glance what the poster is about?

2. Does the poster provide a simple **summary** of its main point?

3. Does the poster **background** let text and other foreground elements be seen clearly?

4. Are the **colors** consistent?

5. Are the **graphs** clearly explained?

6. Is the **size** of the text and graphics large?

7. Are the images **crisp** and not pixelated?

8. Does the poster have enough **white space**?

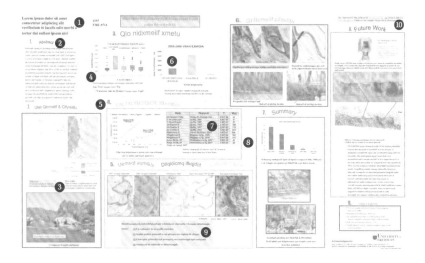

FIGURE 21.2

Problem areas on conference poster.

9. Is all the **text** legible and consistent?

10. Is everything aligned to a **grid**?

11. Does the order follow the expected reading **flow** (top to bottom, left to right)?

12. Are the biggest things on the poster the most important things (**visual hierarchy**)?

13. Are related things grouped together (**proximity**)?

14. Are conclusions supported by **analysis**?

15. Are there **references** to relevant work elsewhere?

If your score is below eight, consider going back and doing a major overhaul of your poster. If your score is between eight to twelve, you probably won't embarrass yourself by putting this poster up. If your score is over twelve, you might have one of the nicer looking posters at the conference. But don't worry if you don't score fifteen. You may have good reasons doing something that this checklist can't account for. Besides, you always want to leave yourself room to improve. Only deities are perfect.

Here's how the example in FIGURE 21.1 might rate:

1. *Does the poster have an easy entry point that lets someone know at a glance what the poster is about?* Arguable, but there are enough photos on the poster to know this is about rivers. Yes.

2. *Does the poster provide a simple summary of its main point?* The poster does have a "conclusion" box in the expected location, but it has five points instead of one. No.

3. *Does the poster background let text and other foreground elements be seen clearly?* The background is plain white. Yes.

4. *Are the colors consistent?* No.

5. *Are the graphs clearly explained?* Yes.

6. *Is the size of the text and graphics large?* No.

7. *Are the images crisp and not pixelated?* This is hard to tell at the reduced size here, but some images are pixelated. No.

8. *Does the poster have enough white space?* No.

9. *Is all the text legible and consistent?* Most of the text can be read, but some is small or distorted. No.

10. *Is everything aligned to a grid?* No.

11. *Does the order follow the expected reading flow (top to bottom, left to right)?* Yes.

12. *Are the biggest things on the poster the most important things (visual hierarchy)?* No.

13. *Are related things grouped together (proximity)?* The boxes do group content together. Yes.

14. *Are conclusions supported by analysis?* Yes.

15. *Are there references to relevant work elsewhere?* A single citation only provides authors' surnames and a year, making it difficult or impossible to track the reference down to the source. No.

The poster's final score is six out of a possible fifteen, with a few points coming more from its content rather than the graphic design.

FIGURE 21.3 shows another sample poster.

1. *Does the poster have an easy entry point that lets someone know at a glance what the poster is about?* Yes.

2. *Does the poster provide a simple summary of its main point?* Yes.

3. *Does the poster background let text and other foreground elements be seen clearly?* Yes.

4. *Are the colors consistent?* Yes.

5. *Are the graphs clearly explained?* Yes.

6. *Is the size of the text and graphics large?* Yes.

7. *Are the images crisp and not pixelated?* Yes.

8. *Does the poster have enough white space?* Yes.

9. *Is all the text legible and consistent?* Yes.

10. *Is everything aligned to a grid?* Yes.

11. *Does the order follow the expected reading flow (top to bottom, left to right)?* Yes.

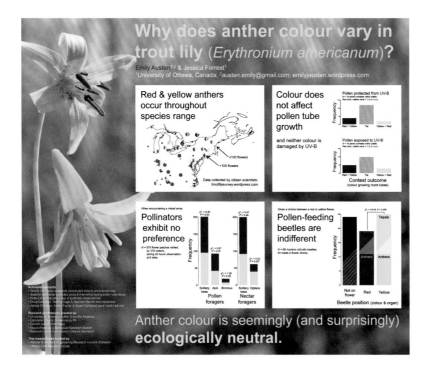

FIGURE 21.3

Sample poster.
(Emily Austen)

12. *Are the biggest things on the poster the most important things (visual hierarchy)?* Yes.

13. *Are related things grouped together (proximity)?* Yes.

14. *Are conclusions supported by analysis?* Yes.

15. *Are there references to relevant work elsewhere?* No.

This poster scores fourteen out of fifteen. This is consistent with the uniformly positive response this poster receives when I share it with readers and audiences.

As mentioned in Chapter 5 (under *Read the instructions*), different conferences may have their own scoring sheets or checklists. You can use those to assess your poster, too, keeping in mind that most focus much more on the content of the poster than the graphics.

Chapter recap

- If you can't read a letter-sized version of your poster at arm's length, it's too small.

- Use a checklist to help you and others review your work.

CHAPTER 22

Printing

Printing a poster requires planning, not least of which concerns who will print the poster. Many institutions have their own large-format printers. There are many commercial businesses that perform large-format printing. If none is available locally, online businesses can often print and ship the completed poster by courier overnight.

It is worth finding a printer early, because different printers have different expectations for how work is delivered to them (preferred file formats, for example) and different turnaround times to finish the work.

Paper

Most printers that work in large formats like conference posters offer several kinds of paper. Matte paper absorbs ink and scatters reflected light slightly, so there are no reflections and glints. Reflections can obscure content, so matte paper is the most accessible to people with visual impairments. Similarly, because 3-D anaglyphs rely on the same shapes being in slightly different locations, glare interferes with the 3-D, so matte paper works best. In contrast, ink stays on the surface of glossy paper rather than seeping into it. Glossy paper tends to have deeper, richer colors. Fine details look sharper than on matte paper. Satin papers are intermediate between matte and gloss.

Some printers may offer the option of laminating a poster. Lamination can help protect a poster against nicks, abrasions, and fading, so may be worth the extra costs if you want to use a poster more than once. Like paper, lamination can be done in matte or glossy finishes. Some glossy laminations allow you to write on them with dry erase markers. Being able to make on-the-spot annotations that you can then erase can make for a more dynamic presentation.

Because most people make posters only with the intention of hanging them up for a few hours at a conference, it is tempting to use the lowest-cost printing possible. But cheap printing often looks cheap.

If you are interested in keeping your poster looking good for a long time, some papers, inks, and dyes are more "fade-resistant" than others. Matte paper tends to hold color longer than glossy paper. Pigments are more fade-resistant than dyes. Some papers and inks are called "archival," but this is an advertising term that has no specific meaning consumers can rely on. Regardless, even a poster printed using the optimum combination of paper, ink, and lamination will not keep its brightness if it is consistently exposed to direct sunlight.

Fabric

Some printers can print posters on fabrics. The kinds of fabrics used, and the technologies used to print on them, are still developing and changing. Different printers use different kinds of fabrics. Just as different paper types change the look of the poster, the final print quality can vary significantly depending on the fabric used. Some are softer and more pliable, like tee-shirt fabrics, while other fabrics are slightly stiffer.

The major selling point of printing a poster on fabric is portability: your poster can be folded so it will fit easily in a suitcase. They downside is that because they are folded, they will wrinkle as surely as the clothes in your suitcase do. Some fabric posters can be lightly pressed to remove the wrinkles after being stowed in luggage. (If you're lucky, there may be an iron in your hotel room.) But even if you iron your poster and get the wrinkles out, you might still notice it sagging slightly after you hang it up, because fabrics are not as stiff as thick paper (FIGURE 22.1).

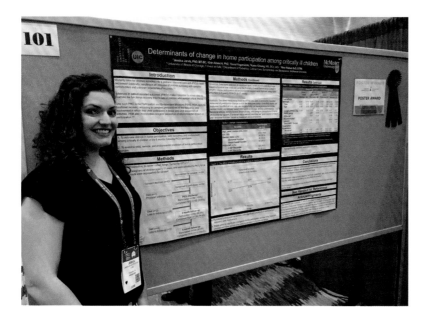

FIGURE 22.1

Fabric poster. (ACRM-Rehabilitation)

The colors on many fabrics tend not to be as bright as on paper. Because fabrics are porous, the inks tend to spread slightly more than on paper and the images may look less crisp compared to the equivalent images on paper.

Checking for errors

There is always a risk that a printed poster will not look the way you expected. But printing is so near the end of the poster-making process that sometimes people do not leave enough time for corrections. You should try to print your poster at least a day or two before you leave for your conference so that you can identify any printing problems and have time to fix them.

Printers vary in their flexibility, their workflows, and how much experience they have of working with academic posters. For example, some businesses prefer a non-editable file format like PDF or EPS. Others, particularly businesses that handle many academic posters, prefer working from the original editable file such as PowerPoint or Illustrator so that they can fix errors directly, without having to get entirely new files from the poster creator.

As mentioned in Chapter 5, the most common printing problem is that the poster is the wrong size. This is a problem that can and should be caught well before the printing. But not all printing problems can be caught in advance. In some cases, you just have to print it and see.

If your poster depends on precise colors, it is even more important to print the poster early so you can review it, because printed colors on your poster may not look exactly the way you expected from the screen. There are several possible reasons for this.

First, as mentioned in Chapter 10, light on computer screens and inks on paper make colors in fundamentally different ways. White spaces on computers are created by shoving photons out of a screen. This means whites on computer screens are literally glowing, so look larger and brighter, particularly if you tend to view the screen in places that aren't brightly lit. The opposite is true on paper. White spaces on paper can only reflect ambient light. Also, ink on paper can "bleed" outwards from where the printer puts it, so black spaces on paper may look larger than on a computer screen, particularly if the ink is printed on low-quality paper or a porous fabric.

Another common surprise at the printing stage is that fonts do not print correctly. For example, not all TrueType fonts can be embedded in PowerPoint, because font creators can set different options for their fonts, including "non-embeddable," "preview/print," "editable," and "installable." Consequently, there is no guarantee that a file that looks one way on your computer will look the same on another computer, and it's difficult to know if there will be a problem until you try it.

The PDF format is intended to preserve appearances exactly as intended. Most software now offers an "export to PDF" option, but strange and unexpected font conversions still happen sometimes (FIGURE 22.2). Once created and checked, however, a PDF file should stay stable and print correctly across different computers.

Some problems are caused by mechanical problems. For example, "banding" happens when a printer is low on ink or dye (Dabner *et al.* 2010), creating light streaks on the paper where there should be solid, continuous colors. Similarly, if the individual primary colors are not aligned properly, there will be "ghosting" around the edges of images. These problems are not the poster maker's fault, but a failure of the operator to maintain the machinery and inspect the work. A poster with banding or misaligned colors should be reprinted for free.

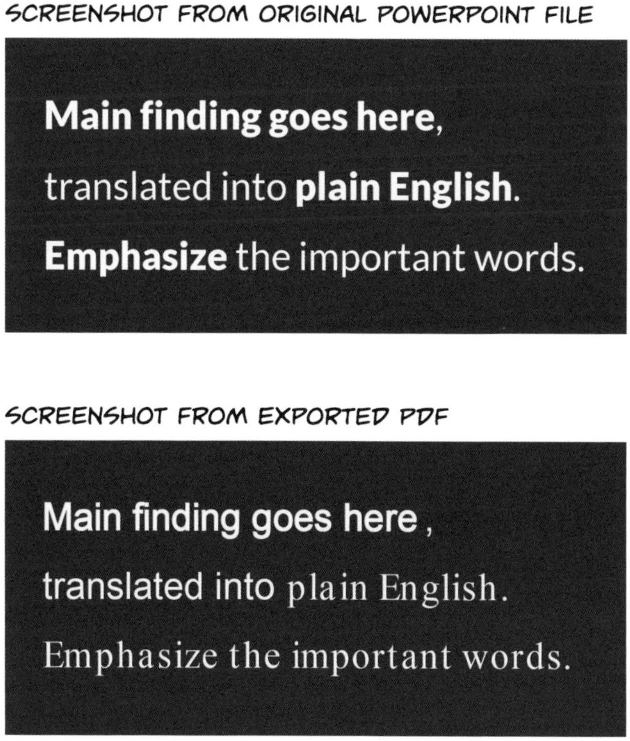

SCREENSHOT FROM ORIGINAL POWERPOINT FILE

FIGURE 22.2

Conversion error causing unexpected typeface changes.

SCREENSHOT FROM EXPORTED PDF

Handouts

When you are having your poster printed is a good time to print any other materials you might want to give out.

Business cards are cheap, easy, and expected networking tools. They are useful to have throughout the conference, not just during the poster session. You will often meet people early in the conference but have a poster presentation days later. A card is a simple way to invite people to swing by your poster.

While most business cards are generic, you might want to put in the effort to make cards tailored to a single conference. If you plan, you can print the title, board number, and presentation time on one side of your card to remind people when to come visit. Some people put representative graphs or images from their poster on one side. The main problem in doing this is that professional print shops often have minimum orders of hundreds of cards, which is usually too much for a single conference.

You can make your own small-batch artisanal business cards for single conferences by using "do it yourself" business-card paper. This is a heavy-weight perforated paper that fits into standard desktop printers. There are standard templates to get the margins right. Usually, a single page will let you print off ten copies of a customized business card, which is the right amount for most conferences.

You can make letter-sized (or A4) copies of your poster to give to interested people as handouts. Most software allows you to take your image and force it to fit a page of a given size. If you have followed the advice in this book, your poster should be readable when it's shrunk down to a single piece of letter-sized paper.

Some conferences explicitly ask people not to bring handouts in the interest of reducing paper waste (International Society for Traumatic Stress Studies 2019). The usefulness of handouts has declined a little, because so many people have smartphones with them all the time. They can just take photos of the poster if they are interested in it. Nevertheless, a handout acts as a cue for people that tells them, "It's okay, I want to share this with you." And some people still like the physical trace.

If you have already published research related to your poster, you might want to print some copies to give to interested people. This can be particularly helpful to people if the papers were not published in an open-access journal. Even if the papers are open-access, giving that physical copy means that the poster viewer is less likely to forget your project, and more likely to read it on the flight home.

Chapter recap

- The look and durability of a poster is affected by choices made at the printing stage.

- Fabric posters are more portable than paper but have slightly lower resolution. And they wrinkle and sag.

- Leave enough time to check the poster so that it can be reprinted if errors in printing happen.

CHAPTER 23

Travel

With your poster printed, you can relax on your way to the conference. You can permit yourself to feel slightly smug that you have no more preparation to do while you watch other people going to the conference still fiddling with their slide decks.

Document tubes

You've printed your big, beautiful poster and are ready to take it to the conference. What you want to get it there is an item is known by several names. Sometimes it's called a document tube, sometimes a blueprint tube, and sometimes a portfolio tube. Whatever the name, these wonderful little items seem to be a surprisingly rare commodity, if the number of times I've loaned mine to colleagues is any indication.

Anyone who presents at conferences regularly should buy one of their own. These are often found in specialty shops catering to graphic designers and artists. Check that the length of the tube exceeds the width of your poster. For example, many nice tubes are only 37 inches (940 mm) long, while many large-format printers can make posters 42 inches (1,065 mm) wide or more.

These tubes rarely have spots to include identifying information, so find a way to attach your name and contact information on the tube. And maybe stick a business card or two inside. When you get on a plane traveling to a major conference, several other people may have tubes like yours, and it would be a shame to walk away with someone else's poster.

Because poster tubes are not typical carry-on luggage, the tubes can signal who else is going to the same meeting you are. When you see someone with a poster tube at the same airport boarding gate, use that opportunity. Ask them, "Are you going to this conference, too?" Remember, the poster that starts the most conversations wins. And look at that, you're having one of those valuable hallway conversations even before the conference starts.

Chapter recap

- Get yourself a document tube.

CHAPTER 24

Networking and presentation

You have made it to the final lap, where you have the chance to network and get feedback and ideas from other conference goers.

Networking and promotion

No matter how beautiful your poster may be, do not believe that all it takes for you to ensure a large interested audience is to hang it up. Be willing to promote it.

Promoting your poster can start before the conference even begins. Some conferences offer "invitations" that allow you to alert people who you think might be interested in your poster. And there is nothing wrong with contacting people you would like to meet by email or social media and asking if they will be at the conference and have a chance to see your poster.

If you do this, the temptation may be to target people in your field who are "famous." It can be fun to talk to academics who are well known. Because academic "fame" is a very limited thing (unlike movie or television fame), most are very approachable and genuinely interested to see posters. But people who are "famous" might already have many commitments during a meeting. The best-known individuals in a field are well known usually because they are well along in their careers, and they might be "conferenced out" generally. After all, some have been going to conferences for decades (Simon 2019). Some may not be active researchers in the field for much longer. But if you are an early-career researcher, you might be in the field for a long time. It is valuable to find people close to your career point. You can see those people again and again, year after year. Those people can become long-term collaborators and allies.

Conferences aren't just an opportunity to build a network. They are an opportunity to gain something more valuable: friends (Simon 2019).

Once you get to the conference, how you promote your poster will depend on the timing of the poster session. If your poster is on the first day, you will have fewer chance to promote it than if it is later in the conference. On the plus side, conference attendees are more likely to be excited and energized on the first day than the last and eager to get into the poster session, so there may be less need to promote.

If your poster is not on the first day, you will have more chances to network. Many extroverts love networking, and many introverts find it exhausting. But talking to other people at conferences is one of the best ways to ensure that some people show up to your poster. It can be as simple as ending a good conversation by inviting people to your poster and giving them your business card.

Social media is a powerful way to network during conferences (Bik and Goldstein 2013). The following discussion refers mostly to Twitter, because it is the social media platform that has added the most value for conferences because of its brevity and the ease of following topics through hashtags (Wilkinson *et al.* 2015, Attai *et al.* 2016).

Promoting your poster on Twitter is an excellent way both to meet other attendees at the conference and to expand your reach beyond the people that you personally meet there (FIGURE 24.1). But tweeting about your poster should only be one small part of your conference experience on social media.

The first word of social media is "social." Simply tweeting, "I'm giving a poster at this board at this time" – and nothing else during a conference – isn't social. It's egotistical. Online conversations are real conversations, and what makes someone a good conversationalist face-to-face holds true in online environments, too. You don't want to be among those people who only talk about themselves. Conversely, don't just lurk and never post. Listening, responding, and starting your own topics all make conversation work and identify you as an active participant in your professional community.

Writing good tweets from a conference provides a service to your professional community. Many people cannot attend conferences that they are interested in (Shiffman 2012, 2017). They want to know what is going on and specifically follow conference hashtags to follow events and talks being given at the conference. Having a built-in audience for tweets on a topic of common interest is an excellent way to build engagement with new people – both with other conference attendees and with people not at the meeting (Bik and Goldstein 2013, Bombaci *et al.* 2016). In one case, 90% of engagement from conference tweets were from people not attending the conference (Shiffman 2012).

FIGURE 24.1

Examples of tweets promoting posters during sessions at the 2020 meeting of the Society for Integrative and Comparative Biology.

Amy Cheu
@BasiliskosArt

Did you there's ANOTHER lizard that can RUN ON WATER aside from the Basilisk? Check out the posters from my lab on Anolis running on water! (P-131 and P-134). #SICB2020

3:01 PM · Jan 6, 2020 · Twitter for iPhone

Natalie Hamilton
@batalie_natalie

Come learn about borrowing sea anemone at my poster!! 24 at 3:30 #SICB2020

7:53 AM · Jan 4, 2020 · Twitter Web App

Julia Notar
@indy_sea

Hey #SICB2020, come check out my poster today! I'll tell you all about visual acuity in different sea urchins. And the cool urchin spine necklace I'm wearing! P1-107

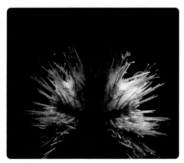

8:54 AM · Jan 4, 2020 · Twitter for iPhone

Kory Evans PhD
@Sternarchella

If you're into skulls, wrasses, burrowing, funny jokes or obscure rap references, come see my talk at 2:15 on Sunday at #SICB2020 !!! and catch me afterwards at my undergraduate's poster for flatfish action!

9:30 AM · Dec 31, 2019 · Twitter Web App

Jennifer Spillane
@jen_spillane

Transcriptome assembly quality affects phylogenetic inference. Come see me at my poster today! (Number 26 at 3:30) #SICB2020

7:49 AM · Jan 4, 2020 · Twitter for iPhone

Laura Bagge, PhD
@laura_bagge

Come by my poster 93 this afternoon to learn about cool optical tricks in scarab beetles! Bonus - you can look at actual beetles through 3D glasses 🪲👓 #SICB2020 @SICBtweets

10:01 AM · Jan 4, 2020 · Twitter for Android

Most academic conferences support social media in some form, although some are more progressive about this than others. Some conferences are *laissez faire*, while some conferences have policies against taking photographs or audio recordings or sharing them. Check the conference website for those policies. But even if the conference policy allows sharing on social media, some individuals do not want their work shared (Pemmaraju *et al.* 2017).

You may also find a conference hashtag on the website or in an email from the organizers. Use this hashtag in all conference-related tweets (Ekins and Perlstein 2014).

You do not have to wait until the conference to start tweeting about the conference. You can tweet about making your poster, traveling to the meeting, who you hope to meet, and other elements that mix both the personal and the professional. You might want to tweet to others going to the meeting to arrange coffee or lunch (Ekins and Perlstein 2014).

While at the conference, live-tweeting sessions as they are occurring is popular and valuable to others (FIGURE 24.2; Shiffman 2019a). It is

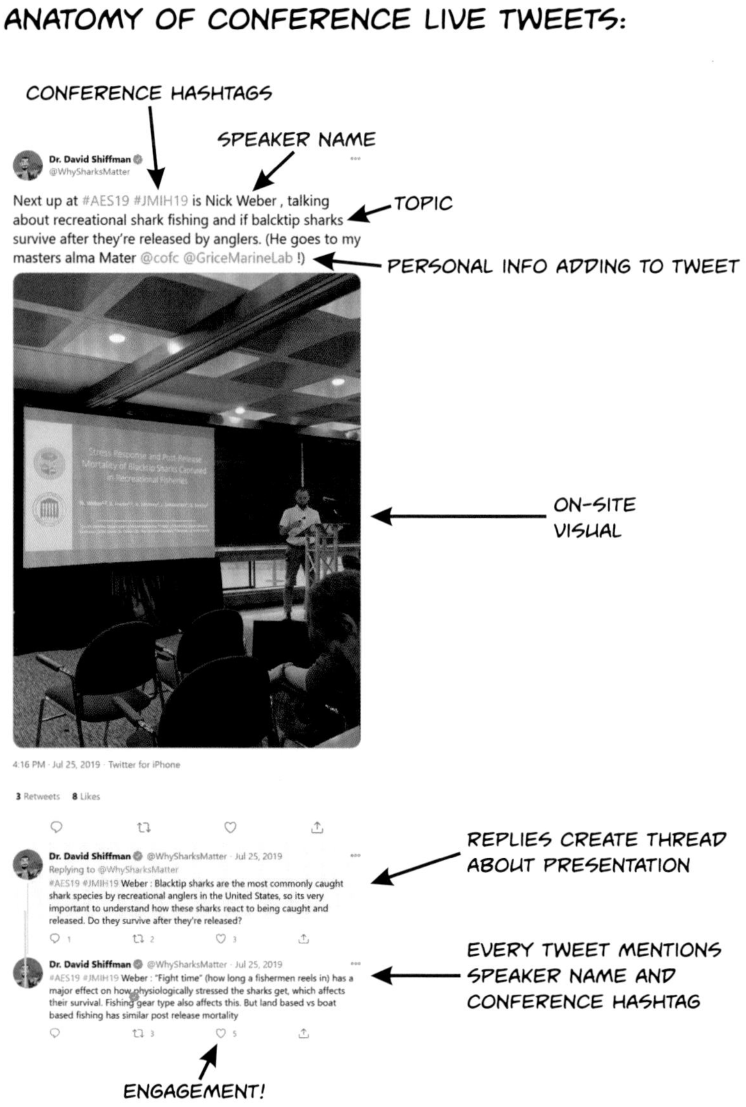

ANATOMY OF CONFERENCE LIVE TWEETS:

CONFERENCE HASHTAGS

SPEAKER NAME

TOPIC

PERSONAL INFO ADDING TO TWEET

ON-SITE VISUAL

REPLIES CREATE THREAD ABOUT PRESENTATION

EVERY TWEET MENTIONS SPEAKER NAME AND CONFERENCE HASHTAG

ENGAGEMENT!

FIGURE 24.2

Example of conference live-tweeting.

good practice to think quickly, write concisely, and build a professional presence (Kinsky and Bruce 2016). There are common practices for doing so (Ekins and Perlstein 2014).

In the first tweet about a talk or poster, identify the presenter (either by Twitter handle or by full name), the topic of the presentation, maybe their affiliation, and use the conference hashtag. In later tweets, you can just start the tweet with the twitter handle or surname of the presenter. (Remember that tweets starting with a Twitter handle normally only show up in that person's timeline, so if you want everyone in your timeline to see the tweet, preface the Twitter handle with a period.)

If the presenter makes multiple points you want to tweet, reply to your first tweet so that the series is threaded together. Include pictures in your tweets whenever possible (Rogers 2014, Cárcamo Ulloa *et al.* 2015).

Don't just tweet presentation titles. That is repetitive and boring. Add your own commentary, as long as you clearly differentiate your opinions from the presenter's (Ekins and Perlstein 2014). Sometimes a talk may turn out to be boring, and you will only end up with a single summary tweet. That's okay. But remember that if you tweet "This talk was boring," be ready to own that.

If you live-tweet a talk so quickly that the tweets are seen before the talk ends, people on Twitter may ask questions about it. Relay those questions to the speaker if you can, making it clear that they are from people online rather than yourself (Ekins and Perlstein 2014).

You will learn quickly if your live-tweeting is on point by checking your retweets and likes. Doing a good job live-tweeting other people's sessions earns good will from conference goers so they will not judge you as purely self-interested when you promote your poster. Tweet your poster presentation day, time, and location both before and during the session. Do not be afraid to repeat it a few times, since people dive into and out of their Twitter feeds, provided it does not make up most of your feed.

If you are in a lab, cross-promotion is a useful strategy. If someone in the lab is giving an oral presentation before your poster, ask if they are willing to mention your poster at the end of their talk, particularly if the topic is related to yours. If there are posters from the lab scheduled before yours, or hung in different locations in the conference room, ask if they can mention your poster in a small sidebar.

The presenter(s)

In general, one of the authors of the poster is the main presenter. There are advantages to having a single presenter who can synthesize the questions and feedback that came from everyone who stopped at the poster.

Because the research shown on posters may be the result of collaboration, it is okay for presentations to be collaborative, too. There may be multiple presenters, either simultaneously or consecutively. (That is, one presents for an hour, then another presents for an hour, and so on.) The presenters should not outnumber the audience, however.

It's not recommended that someone who was not involved in the work (that is, a non-author) present a poster (Foster *et al.* 2019), but there may be rare cases where this is unavoidable.

What to wear and bring

The most important things in your conference wardrobe, particularly when presenting a poster, are your shoes. Many large conferences are in convention centers that are hundreds of yards from end to end. You could be standing for hours in front of your poster. It is important that what you wear on your feet is comfortable for that length of time.

Conference clothing can be a vexing issue, particularly if you are going to a conference that you have never been to before. Different conferences can have different levels of formality. For example, medical researchers tend to dress more upscale, while field biologists tend to sport blue jeans and sandals. Furthermore, descriptions of preferred dress provided by the conference are not always helpful. Morgan Simon (2019) noted that someone from Kansas City or San Francisco might have a very different interpretation of "business casual" than someone from New York. If anything, new presenters tend to overshoot the expected level of formality. While it might be good to be dressed a little better than the audience, it's undeniably awkward when someone shows up in a tuxedo or evening gown when everyone else is in tee-shirts.

Besides the level of formality, there are often other common practices within the field. Anthropologists heading to conferences often wear jewelry representative of the cultures they study (and have a thing for scarves) (McGranahan 2013).

If you have never been to this conference before, ask others who have been what the atmosphere is like. Look online for photos from the previous conferences, both on the organization's website and on social media.

Because many conferences tend to be "male and pale," conference attire can be even more complex for anyone who doesn't fit that mold. For example, many buildings are temperature-controlled under the assumption that all its occupants will be wearing suits, meaning it can be cold for anyone not so dressed (Simon 2019). Many women rightfully complain about the difficulties of choosing conference clothing (Renee 2018). Femininity is often discouraged, particularly in the sciences (McSweeney 2019). "For women, clothing requirements are ambiguous and options seemingly infinite. And, as much as we hate it, women can still be judged harshly for their looks in a male-dominated scientific community. Even by other women!" (KK 2012). Holly Renee (2018) advised, "Acknowledge that most of the rules that exist for women's professional dress for a conference are steeped in misogyny and outmoded standards of female purity."

What almost everyone agrees on is that regardless of whether you wear a tee-shirt and jeans, or crisp white blouse and dress slacks, you can almost instantly blend into an academic setting by putting on a blazer, jacket, or sports coat. Tweed and elbow patches are helpful, but not required.

But there can be advantages to standing out instead of blending in. If you look different than most conference goers, Morgan Simon (2019) suggests embracing that difference: "I learned as a 23 year-old that I was going to look like I was 15 if I showed up in a traditional western-style suit." Distinctive apparel can be a networking tool, for several reasons (Rohde 2019, Simon 2019). First, it makes you recognizable. If you have been told, "Look for the man in the kilt," the chances are good that you will recognize the person you're looking for if your paths cross, even at the largest conferences.

Second, it makes you approachable and starts conversations (Renee 2018). Wearing something unusual gives people the chance to talk about something different than the routine introductions of university and field of study. "Why are you wearing a kilt? Are you Scottish?" "No. It's my reminder to myself to be fearless. Because a man in a kilt fears nothing."

Third, a single distinctive item can tell people something about you. Colorful hair tells people you are not afraid to be a little unorthodox. A cowboy hat might signal that you're from the western part of America.

Charles Cong Xu took this philosophy to the next level at a joint meeting of the Canadian Society for Ecology and Evolution, the Entomological Society of Canada, and the Acadian Entomological Society in 2019 (personal communication). Xu's work was on spider webs, so he decided to use one of the most familiar ambassadors for spiders in popular culture, the superhero Spider-Man. First, Xu used some of the visual elements of comics to inform his poster. While presenting, he upped the ante by wearing a partial Spider-Man costume (FIGURE 24.3).

FIGURE 24.3

Charles Cong Xu presenting his poster. (Charles Cong Xu)

"Go big or go home" was why I decided to dress up and have some fun with it. The socks and compression shirt worked well, but the mask was stuffy. I ended up just putting it on every once in a while when people wanted to take photos.

I did not get any pushback about the costume. On the contrary, the poster and costume drew a lot of positive attention at the conference, as Twitter would testify … I think it's a good sign when people want to take selfies with you and your poster.

Not everyone needs this level of commitment to their theme, however.

Whatever the level of formality, wear what makes you feel confident. Some people want to wear their "power suit" when they present. That's what makes them feel confident. Other people despise suits and ties and it would be a disaster for those people to present in them. Check how your clothes look from multiple angles to ensure you are comfortable. Morgan Simon (2019) wrote:

> I learned … that I often was coming off as sexier than I had intended – I wasn't checking how far my skirt was riding up when I crossed my legs, or how deep a v-neck really was when standing from above, until later seeing these things on video and actually being surprised that I was showing more than I had intended.

One paper suggested you should wear clothes that match the colors of your poster (Keegan and Bannister 2003), but this conclusion is doubtful. For one, the sample was tiny: just two presenters at two posters at one meeting. Color-coordinating your clothes with your poster runs the risk of looking gimmicky.

Besides the clothes on your back, there are other items you may want to have with you. You could be standing for hours in front of your poster, talking almost continuously. You may get thirsty or hungry, so stash a drink to keep your voice clear and bring a snack to keep your energy levels up.

Pointers can be useful. Telescoping physical pointers, about the size of a pen, can be found in office stores. Some presenters use laser pointers that are normally meant for talks.

If you have supplemental video as part of your poster, take a laptop or tablet that you can use to show your video.

Hanging the poster

Poster boards are often available for hanging well before the poster session starts. Sometimes, the poster boards are up and available first thing in the morning, but poster sessions are not until mid-afternoon or later. Hang your poster up sooner rather than later. People who want a break from the conference will often walk through the aisles during coffee breaks and lunch. Some people prefer to read posters quietly. Some people look for posters to come back to during the session. Either way, you are missing opportunities if you only put your poster up at the start of the session.

Try to get the poster up as high as possible so that the title can be seen over the tops of peoples' heads. This might require asking some bystanders for help. If the top of your poster is not almost flush with the poster board, you have made your poster too small or you are mounting it too low.

When you are hanging it, ask someone else to tell you if it is straight. I can never tell when I am hanging it myself and always end up with one corner higher than the other.

In theory, four tacks should be enough to hold a poster in place. In practice, paper posters often need a couple of extra tacks along the middle to stop the paper from buckling. Fabric posters may need even more tacks along the top to minimize sagging.

The conversation

Poster presenters fall somewhere along a continuum between aloof and desperate. You look aloof if you are looking at your phone or tablet for more than a few moments, sitting on the floor reading or eating, or engrossed in talking to someone else. You look desperate if you stare at people with puppy-dog eyes as they walk by your poster or randomly grab people who walk even slightly near the area of your poster and launch into your spiel. These tactics are not likely to start conversations about your poster.

The mid-point between those two extremes is being approachable and inviting. When someone walks near your poster, step up and say, "Hello! This is my poster. Let me know if you have questions, or I can give you a quick run-through if you would like." Be willing to take "no" for an answer if someone says, "I'd just like to read it, thanks." It might feel a little strange watching someone read your poster while you're right there, but you need to respect people's preferences.

If someone says, "Yes, please tell me about the poster," a good way to start the conversation is to say your ABT sentence (Chapter 6), then take a moment to ask about them (Brookshire 2019). Find out why they stopped at your poster. The chances are good that they are interested in similar problems – otherwise, they wouldn't be at your poster! Data can be a deeply personal experience. People search for connections between themselves and your data (Serrell 2015, Peck *et al.* 2019). For instance, if your poster shows a map of Australia, you are very likely to get Australians stopping at your poster, or people who have lived in Australia, or people who have visited Australia.

Then, try to find a way to connect your project to their interests and build off it (Olson 2016). This technique makes guided tours more enjoyable and informative (Tsybulskaya and Camhi 2009). Then, ask how much time they'd like to spend. Would they like a one-minute summary, or a longer (and more detailed) five-minute tour? (Keep in mind that your "five-minute tour" should only have four minutes of you talking.)

While engaging with an audience member is easy to say, Nina Simon (2016) notes, "it's not easy to retrain yourself if you've spent years rattling off your offerings to each new person who approaches the desk. It's not easy to stop, ask, and listen before you speak." Listen with the same intensity as you talk (Wurman 2001).

You may want to briefly introduce yourself, focusing on your organization and role more than your career stage (Simon 2019). It can help prevent the awkward social situation of people assuming you are more senior or junior than you are. When you are a faculty member, it's annoying to be mistaken for a graduate student. Make your knowledge the focal point instead of your age.

When you were laying out your poster, you should have developed an idea of how you would explain your poster, and perhaps even sketched out a script. Don't "can" your speech. It shouldn't be memorized like a one-act play. Be ready to do a short version that takes a couple of minutes, and a longer version for people who want more detail.

When explaining graphs, first describe what the horizontal and vertical axes show, in that order. Then describe the inside data, and finally interpret what the data mean (Roam 2013).

When you are talking to one person, keep an eye open for other people who walk up. Take a moment to reach out to say hello to them, so they don't feel ignored and walk off. They might never come back.

You can quite often get several people listening to you talk about the poster at once. Some shy audience members strongly prefer listening this way! They want to hover around the edges with a few other people. Because of this, you might try to start your session by asking a friend if you can practice your poster talk on them. Having someone there lowers the intimidation factor for other people.

One of the problems an audience member can face is wanting to leave. Maybe they realized that the topic is less relevant to them than they hoped. Maybe they need to pee. It doesn't matter why they want to leave, but most people will want to be polite and let you finish your presentation. There is no simple way for people to extricate themselves from someone giving a tour through the poster. This is one reason some people like to "eavesdrop" on a poster presentation: if you're one of the crowd, you can quietly slink away.

If someone looks distracted, or just not into it, offer them an "out." Ask, "Is that sufficient?" or "Would you like to know a little more?"

Many scientific conferences are in English. If English is not the language you feel most comfortable speaking, there are resources on presenting for non-English speakers (Wallwork 2016, Hartley and Sheridan 2019).

Inclusive presenting

Poster sessions can be busy and complex environments, and not everyone has the same needs.

Some audience members might have mobility devices or service animals to assist them. Make sure to keep your tacks and pins off the floor, where they can potentially puncture tires or poke the paws of service animals. Do not touch their devices or pet service animals without asking permission, no matter how cute the animal may be. Service animals are working, and petting is as much an unwelcome distraction as if someone came into your office in the middle of the work day and started hugging you for no reason (Society for the Study of Evolution 2019).

Audience members who do not hear well or do not speak the same language as you may have an interpreter. The interpreter may be a friend or acquaintance of the viewer, not a dedicated professional, so may not introduce him or herself as an interpreter. Regardless, speak to the person reading the poster or asking the question, not necessarily the interpreter. (If the interpreter asks a question, then talk to that person, by all means!) If a person has difficulties hearing but no interpreter (which may not be obvious from looking), face towards them so that they can see your lips, and don't talk while eating or chewing. If another audience member asks a question, repeat it before answering. Offer to meet at a different location that may be less noisy.

If there are audience members with visual impairments, introduce yourself at the start. Remember, they may not be able to read your name tag! As you talk about the poster, remember to describe all the content on your poster, including the charts, graphs, and pictures. If guiding someone with a visual impairment, offer them your arm rather than grabbing or pushing them.

Exit strategies

The benefit of poster sessions, the one-to-one interactions between presenter and audience members, is a double-edged sword. You never know who will show up. Sometimes it's a person you don't want to see.

Sometimes it's a person who overstays his or her welcome.

Sometimes it's a bully. Rebecca Shansky (2010a) described having a "science enemy" who harassed her colleagues, culminating in this move: "[She] once came up to my poster, whipped out a ruler, and proceeded to *measure my error bars*. Who *does* that??"

And sometimes it's just a creep. There are, unfortunately, people who do not understand that a poster session is a place to exchange ideas, not phone numbers. Far too many poster presenters, mainly women, have experienced men being creepy to them. To put some exact numbers to this, over 1,100 women responding to an online survey reported that they had been made to feel uncomfortable by a man who wouldn't leave their poster 25–100% of the time they presented. Sixteen percent of women said that this happened to them at every conference (McLaughlin 2018a).

Uncomfortable as those numbers are, some encounters women had at poster sessions that they later described on social media are even less comfortable.

> A PhD student … comes to my poster at EVERY SINGLE LOCAL CONFERENCE for 30+ minutes, putting off everyone else and freaking me out. Some of his highlights include, "I like your dress." I now get male friends to look out for him and warn me. (Hockin 2018)

> My fav was the time the fella didn't ask any questions about my work, but did let me know he saw me swimming at the hotel pool earlier. (Harrison 2018)

> A guy spent 10+ min at poster, complimented work, asked if he could call me. My reply: "Uhh, email is the preferred method of contact." Looked down and was alarmed to see MY number going into his phone – he lifted it from my poster tube's "if found call" info. (MacNamee 2019)

> A significantly older guy [spent] half an hour at my poster, no background in my field (i.e., asked what the words in the title meant, and what the differences between flies and rodents are). Kept touching my back/shoulder and would not move on until he took a photo with me. Still … bleh. (Creed 2018)

> Called me sweetheart, while trying to show me he knew more about stats than me. Leered for an excessive amount of time. Blegh. (Sheffield 2018)

These stories are terrible on so many levels, and it is shameful that research communities have been very slow in recognizing these problems, never mind addressing them. For one meeting, 14% of respondents reported receiving unwelcome behavior (O'Meara *et al.* 2019). Bars and socials were the most common places where unwelcome behavior occurred, but poster sessions ranked third as a place for such encounters.

Organizers have a role to play in setting expectations for conference attendees, but presenters should have their own plans in the event of unwelcome guests. Perhaps the easiest is to make sure the poster presenter has a contact number of someone they can text if they want support.

This may be a supervisor, a fellow student, or a conference buddy. The presenter's support team should walk by from time to time to make sure everything is okay. Coauthors and supervisors can often get a student out of a tight spot by asking the presenter if he or she wants someone else to take over presenting the poster for a while. It can be helpful to set up a code for trouble, either by text or verbally. Beth Ann McLaughlin suggests asking the poster presenter if they need a red marker. Replying "yes" means the presenter wants help (McLaughlin 2018b).

Helping a presenter can sometimes be as simple as joining the conversation to give the presenter someone else to talk to. If you are beside your poster and see someone looking uncomfortable nearby, walk over, particularly if the presenter is alone.

Sometimes, it might require a supervisor asking the unwanted visitor, "You've been here a while. What do you need?" If the person says they are chatting, ask them to move on so that the presenter can work. Particularly bad behavior, like unwanted touching, might run foul of a conference code of conduct and should be reported.

Every poster presenter should feel someone has their back if they need it.

As an audience member, be thoughtful about what you might do that might make someone feel uncomfortable. Large conferences are wonderful opportunities to meet people from around the world, but people from different cultures may have different expectations about what is considered appropriate professional interaction or personal space. Some people may be more comfortable with a handshake or a kiss on the cheek than others. But you should never use "culture" as an excuse for physical contact. You can avoid many potential problems by asking permission first.

Chapter recap

- "If you hang it, they will come" is not a viable networking strategy. Promote your poster.

- Wear comfortable shoes.

- Your presentation should not be a canned spiel, but responsive to the audience member you are talking to.

- There can be unpleasant people at conferences. Have a plan to extricate yourself from their presence.

CHAPTER 25

After the conference

Posters are generally created as single-purpose documents. But there are many creative ways to get extra mileage out of your poster.

Encore presentations

Once you have presented a poster at one conference, you may be tempted to present it again at another conference. Conferences rarely have specific rules prohibiting this. If there are, respect those rules (Foster *et al.* 2019). Even if there are no rules forbidding you to reuse a poster, it may not be a good look. If you are presenting to separate conferences in a short span of time (maybe less than a couple of months apart), the reuse of a poster is probably acceptable. But read the instructions for authors for the new conference. If the poster boards are a different size, your old poster might be too big for the new meeting, or your poster could be so small compared to the new board size that it looks ridiculous.

It is poor form to stick up the exact same poster multiple times at different conferences, particularly if they are at the same venue (Foster *et al.* 2019). Most academic conferences are held one year apart, and in the space of a year, you are generally expected to have made some progress on a project: more data, new insights. A doctoral student might present on some project at several meetings in a row, and similar posters are not only accepted but expected. But presenting the exact same poster twice screams that you have made no progress.

If you have a poster from a previous conference, you could use that as the starting point for a revision for the next conferences. If you have saved your original files, it should be easy to revise the title and color scheme and add some new data and text to make a poster specifically for each conference. While this can be more expensive than reusing a poster, particularly if you must pay for large-format printing, the positive impression created by a fresh look is often worth the cost.

If you do reuse a poster, with no new data or changes, be transparent about it. Include a sentence in the abstract that this work has been presented before (Foster *et al.* 2019).

Hallways

Putting up posters in department hallways is a common fate for many conference posters. This makes hallways more interesting to look at, gives students something to read while waiting for professors, helps promote work to prospective students, and lets colleagues in the department know what you've been up to lately.

Hallways do not always have poster-board material everywhere where people can tack up posters. Building maintenance tends to disapprove of sticking tacks in drywall or using masking tape that peels off paint. Some adhesive putties can also stain paint. Mounting strips (sometimes specifically called "poster strips") are special double-sided tape strips that will not damage paint if they are removed slowly and carefully. These can be bought at many home improvement stores.

To keep your poster looking its best for the longest, try to hang it away from direct sunlight or other bright lights. The less light, the less fading. That's why museums and art galleries are often dimly lit.

Besides your own hallways, you might consider donating them to a local high school (Science Café Little Rock 2019). Researchers in universities often forget that we generally have more resources than teachers, who are often looking for new ways to make their content relevant to students. Many high-school students expect or hope to attend university, but they may not have a clear idea of what projects local scientists do. Showing research done locally could be a powerful way to connect students' studies to research projects.

There are even more possibilities for inspiration if the poster shows research by undergraduate students. This might help high-school students see that making real research contributions need not be in the far future, but something that is potentially right around the corner.

Arts and crafts

If your poster has outlived its usefulness in the hallway, it can be a great outlet for your creativity.

Paper posters can be turned into massive origami projects. More than one poster has been turned into a big paper airplane at the end of a conference, but this is usually just a prelude to the poster going into the recycle bin.

The winter holidays are among the best time to consider reusing your paper posters. Posters can make great gift wrapping for the geek in your life (Sandquist 2014) or can be repurposed into holiday decorations (Shansky 2010b).

1. Cut your poster into six equal-sized squares.

2. Take one square, fold it in half along the diagonal to make a triangle. Fold it in half along the diagonal again to make a smaller triangle (FIGURE 25.1).

3. Hold the triangle so that the fully folded side is on the bottom and the longest side faces the hand you will have your scissors in.

4. Make four evenly spaced cuts parallel to one of the short sides. Do not cut all the way to the other short side!

5. Open the square.

6. Take the two innermost flaps and overlap them to make a hollow tube. Tape them together.

FIGURE 25.1

Steps for turning poster into snowflake.

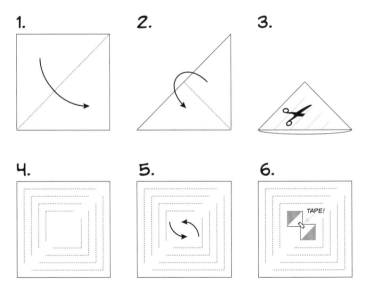

FIGURE 25.2

Poster converted into holiday decoration.

7. Flip the paper over to the other side, take the second innermost two flaps, overlap them slightly, and tape. Keep flipping, overlapping, and taping until finished. One sixth of your snowflake is done!

8. Repeat steps 1–7 with the remaining squares.

9. Once all your snowflake arms are made, arrange them so that one end of each arm is almost touching all the others, in a six-sided snowflake shape. Staple everything together in the middle. Attach the outer arms to the poster board or wall so they aren't floppy.

Following these directions, I turned one of my own posters into a winter-wonderland decoration (**FIGURE 25.2**).

Sandra de Vries was among the first to realize (circa 2014) that fabric posters offer a huge number of possibilities for reuse (Roberts-Artal 2016, de Vries 2017). They can be turned into a scarf or cape with little effort. If you can sew, your data can become anything you can stitch together (**FIGURE 25.3**). Fabric posters have been turned into beach blankets (Werner 2016), tote bags (Roberts-Artal 2016, Hernandez 2017, Lantz 2018), quilts (Stuart 2018), cushions or pillows (Mandy 2015, Cohn 2017), skirts (Roberts-Artal 2016, Cheplygina 2017), suits (Hut 2017), ties (Roberts-Artal 2016), and even a full dress (Kuruwita 2019).

FIGURE 25.3

REPURPOSED FABRIC POSTERS!

Repurposed fabric posters. Upper left: dress made from two fabric conference posters (Rajika (Reggie) Kuruwita); upper right: tote bag (Cheryl Lantz); bottom left: skirt (Veronika Cheplygina); bottom right: pillows (Neil Cohn).

Archiving

Because many posters are not published as journal articles (Scherer *et al.* 2018), archiving them contributes to the scientific record and helps posters reach a much wider audience than just the conference they were presented at.

In a perfect world, conference organizers would archive all abstracts and posters presented at a conference. But many conferences, particularly small ones, do not do this because they don't have the resources and infrastructure. Fortunately, there are now many straightforward options for self-archiving a poster.

One of the most straightforward ways to archive a poster, and to share your conference experience, is to blog about it very soon after the conference. This is an excellent technique for summarizing your experience with the benefit of a good night's sleep and a chance for a little reflection. You can write about what kinds of interactions you had, and where your poster fits into the current trends in the field.

Institutional repositories exist in many universities for all manner of scholarly publications (Crow 2002, Ware 2004). Repositories vary widely in their scope and services according to institutional need and faculty interests (Drake 2004, Abrizah 2017), so some may not archive posters. Some repositories may archive a physical copy of a poster, although more are interested in digital submissions that can be accessible online. Institutional repositories are usually managed by library services, so researchers should contact their librarians for guidance if questions arise.

There are many commercial websites where people can upload their works, including academia.edu, ResearchGate (researchgate.net) and Figshare (figshare.com) (Van Noorden 2014). ResearchGate has several million users, "astonishing for a network that only researchers can join" (Van Noorden 2014). Some, such as Figshare, assign a digital object identifier (DOI) to submitted works, which provides a more stable internet address for long-term reference.

More researchers upload their scholarly output to third-party commercial services than to institutional repositories (Borrego 2017). These websites have traditionally been little moderated and have allowed all sorts of file uploads. But there are serious questions about the long-term stability of these sites. ResearchGate has been sued multiple times for copyright infringement (Chawla 2017, Else 2018). Even if these sites survive the legal challenges, there are also questions about how they might commercialize the information provided by researchers.

You can always archive your poster yourself. For example, you can post it on your own website. You can write about it in a blog post. You can share it on Twitter or other social media. Archiving a poster by yourself has the advantage of simplicity but may make your work harder to discover in the long term than if you put it on one of the better supported "hubs" of scholarly publication.

If you upload your poster online, make the poster accessible. People who find your poster online may be accessing it through a screen reader, not a standard web browser. If your poster is saved as a single static image (e.g., a JPG or PNG file), a screen reader will simply read the title of the file, leaving a viewer completely mystified as to what the poster's content is. You might want to create a "plain text" version of the poster to accompany the file upload.

File formats like PDF are potentially more accessible than a static image, because text is incorporated into the file as text, which can be read by a text-to-speech converter. Many PDF readers have a text-to-speech option built into the software. But there are two potential problems that you need to check.

First, check that the text of your poster can be read correctly by a speech synthesizer. Many little design tweaks to help sighted readers, like hyphenation or kerning, break words into pieces from the point of view of a screen-to-text reader and ruin its ability to read the text of a poster. You may need to go in and undo hyphenation and kerning if you included them.

Second, every image needs an "alt text": descriptions of photos and images for people who cannot see them. Different software packages have different ways of adding alt text. For example, in PowerPoint, right-clicking an image reveals an "edit alt text" option. Once added to the PowerPoint file, the alt text is kept if the file is then exported into a PDF.

Writing alt text is like creating a poster: a balancing act between concision and precision. Phrases like "image of" and "photo of" are usually redundant and can be cut. A description like "dog" is succinct but not as accurate as "golden retriever lying in grass." The optimal length for alt text is a sentence or two, but complex scientific graphics or flowcharts can be exceptionally challenging to shorten.

If the image has any text, the alt text should usually transcribe it exactly. Correct punctuation can also help speech synthesizers sound more natural.

If you have an image that is not conveying any information, it can be marked as "decorative."

If a paper based on work presented in the poster does get published, add a citation to the documentation accompanying the archived poster if possible. That way people who stumble upon the poster can easily find the paper, and can then cite the published paper rather than the poster (Foster *et al.* 2019).

Citations

As with any form of scholarly communication, the litmus test for a conference poster is whether other people use the information the poster presented. The usual way to measure reuse of scholarly product is to look at citations. The distribution of poster citations is undoubtedly skewed, and the mode is undoubtedly zero. But the tail is longer than expected.

The apparent record holder for "most cited poster" is a 1993 poster presenting a post-traumatic stress disorder (PTSD) checklist (PCL) that has been cited over 4,000 times, according to Google Scholar (Weathers *et al.* 1993). The success of this poster was because it solved a problem for people (Ogilvy 1963). Lead author Frank Weathers (personal communication) wrote:

> The PCL is a National Center for PTSD product, so got a lot of visibility. It was also the only validated DSM-correspondent PTSD questionnaires at the time, and there was a huge need in the field for it. It is probably the most widely used PTSD measure to this day. Everyone who uses it in their research needs a citation, and since we never published those specific data, the poster has been the primary citation.

While the checklist has undergone multiple revisions, the National Center for PTSD has continued listing the poster as the primary reference for the checklist (National Center for PTSD 2019).

But finding these highly cited posters relies mostly on word of mouth. Tracking citations to posters is difficult because citations to conference presentations are written in a huge variety of formats that may be difficult to parse. Some abstracts get confused with later publications of the project, particularly when authors do not change the title from conference to final journal article.

The citation counts for poster sessions are driven down because people have differing opinions about whether posters should be cited in journals. When asked, "Are poster session presentations citable in manuscripts?" 50% of respondents said "no" (Shanks 2019). Editors are likewise split. Some editors are fussy about what they will allow in their reference lists and only allow peer-reviewed papers.

Whether posters should be cited or not depends on whether one considers posters ephemera or part of the scientific record (Foster *et al.* 2019). Because many conference presentations never turn into papers (Scherer *et al.* 2018), conference abstracts and posters may be the only record of some claim or experiment.

In the past, many conferences never distributed abstracts beyond paper copies given to attendees. But some conferences have a long tradition of publishing conference abstracts in journals, which meant their poster abstracts were as findable as any journal article. The level of documentation in an abstract may not be as good as in a published paper, but as more conferences accept posters in digital formats and take advantage of high-capacity storage, complete posters should be archived as systematically as abstracts sometimes are. If a poster has no published abstract, it might be better cited as a "personal communication" rather than a presentation, if the journal permits personal communications.

Citing conference abstracts should be not only possible, but something that people do more often for the sake of transparency and history. I've cited posters that presented earlier versions of work in the final manuscript I submitted to a journal: "This work has been published in abstract." This is just the common scholarly practice of citing relevant prior work, and it allows others to track the progress of the work. Because posters are often "work in progress," some of those posters have ideas that turned out to be wrong. The development of ideas can be as useful as the final claims themselves. Our view of the history of the development of the concept of natural selection would be impoverished if we did not have Darwin's copious notes and letters – objects that are usually considered to be as ephemeral as posters.

Even when conference posters do turn into journal articles, they may not turn into journal articles fast enough for a manuscript that someone is working on. Citing a poster or other conference presentation is better than a "personal communication" or unsupported claim. Those citations, with titles and complete author lists, can make it easier to find a final "version of record" if it is ever published in a journal.

Chapter recap

- Don't throw your poster away after the conference. They can be repurposed within your department or made into many different crafty projects.

- Archiving posters online creates a whole new audience for the work and increases the chances of the poster being cited.

PART III

For organizers

CHAPTER 26

Poster session planning

Designing a poster session is, in some ways, like designing a poster: you need to make decisions with a lot of empathy for other people, and not decisions that are just convenient for you. If you want to have a great poster session at your conference, here are things you need to consider.

Format

One of the first decisions conference organizers need to make is whether to hold a poster session at all. Poster sessions are not without their problems and critics (Rowe and Ilic 2015). Some conferences, especially small ones, might simply opt for everyone to give a talk, or promote some other means of networking.

Assuming the conference does want a poster session, there are now options as to whether to do a traditional paper poster session, or an electronic poster session, where posters are displayed on large screens.

Paper is simple, cheap, tried, and tested. But digital posters are an increasingly viable and attractive option for conference organizers. The primary advantage is undoubtedly that they can provide useful data. Organizers can see which posters and sessions were viewed and downloaded, and can then use this to help determine programming in coming years. The benefits for presenters can be an easier way to contact colleagues and share their posters with a few clicks, better time management, and greater access to all posters.

Well-organized digital sessions can schedule presentation slots in such a way that fewer presentations are happening simultaneously. This reduces the competition for attention for presenters and makes it easier for attendees to see more of the presentations they want.

Digital posters can be archived, creating a valuable repository and history of the research in the field.

Submission forms

The way that poster abstracts are submitted can help shape good practices. For instance, forms should try to avoid limiting the number of authors. Allow non-typical contributors, like study groups.

In addition to the title, author names and affiliations, and abstract text, the submission forms could include spaces for ORCID numbers to identify authors, trial registration details, sponsor or funding agency information, and disclosure of medical writing support (Foster *et al.* 2019). These should all be excluded from any word or character limit, and should be published with the abstract.

Another helpful option would be to ask if there are related posters from the same group that the authors would like displayed together.

Poster boards

Poster makers need to know how much space they have on which to place their poster. Once that poster is printed, it can't shrink. It sounds easy: pull out a tape measure and you're done, right?

Many organizers put up numbers on their poster boards so that presenters and audience members can quickly locate their place. But often, organizers practically nail the numbers to the boards, so that presenters can't remove them. And they neglect to account for the space taken up by their solidly attached numbers. Anyone who makes their poster to maximize the space will have their poster not fit – or at least struggle to move the number. If the poster boards are 48 inches (1,220 mm) high, and 4 inches (100 mm) is taken up with the poster numbering, tell people that the maximum poster size is, say, 42 inches (1,065 mm). Leave room for error.

Another mistake that organizers make is to measure the boards from edge to edge, including the board's metal frame. The frame typically takes up an inch or two on all four sides. Posters cannot be tacked into the frame, so again, anyone who followed the instructions would end up with a poster that does not fit in the designated space.

Poster viewing space

Many poster sessions are sabotaged by a single problem: the organizers don't provide enough space for the posters.

Conference organizers faced with a limited space will reduce the maximum poster size, usually to fit two on a board. Yes, you can fit more posters if you have them 4 feet (1.2 m) wide and double up two posters to a board, but you increase the crowding.

Many meetings skimp on space, and the poster sessions are uncomfortable to walk through. They are too crowded, too noisy, and there is only room for the presenter and maybe one or two audience members at the same time. This is particularly an issue for people in wheelchairs or with other mobility issues, and for those with service dogs (Devitz 2019).

A poster should probably have a minimum of 15 square feet (1.4 m²) in front of it, and 20 square feet (1.9 m²) is better. Get the area of the room the session will be in and divide it by the number of posters.

Walkways between posters should be wide to accommodate the presenter, the viewers, and passers-by. It is not reasonable to expect viewers to walk through a poster session in single file (Yoder 2019). Narrow walkway space is a severe accessibility issue for wheelchair users and many others.

Let's work off the notion that a poster is 4 feet (1.2 m) wide. Between the presenter and the audience members, you probably want another 4–5 feet (1.2–1.5 m) of space in front of the poster for people to stand. If two poster boards are in front of each other, that means the boards should be 10 feet (3 m) apart – but that leaves no room for walkways. Another 5–6 feet (1.5–1.8 m) of space – clear of tables or other obstructions – will allow for people to walk in both directions comfortably.

Do not put anything in the walkways, such as tables. Too often, there is a table stuck apparently at random in the middle of a walkway. The apparent rationale for this intrusive furniture is to provide people with a space to socialize, but tables cause congestion in foot traffic and make it harder for people in the session to navigate. Put tables at the end of the poster row.

Do not put poster boards in stairwells or any other place that does not have a flat floor.

Poster boards should run in straight lines, parallel to each other. Placing poster boards at an angle relative to each other causes viewers for one poster to block the viewers of the adjacent poster (FIGURE 26.1).

Poster boards should not be positioned with one edge along a wall. This creates dead ends for pedestrians and congested foot traffic around posters. Because poster content is generally arranged from left to right, posters with their left edges along the wall will have their critical introductory material in a position that makes it hard for passers-by to spot it (FIGURE 26.2). These posters are at a disadvantage in finding an audience.

Do not stack posters on top of each other. This means either that the top poster is well above the head of the presenter, or (worse) that to view the

FIGURE 26.1

Diagonal poster
board placement
forces adjacent
presenters to talk
across each other.

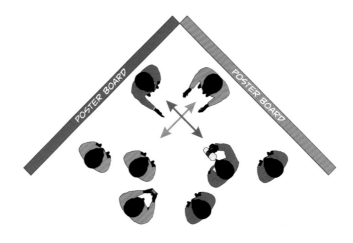

FIGURE 26.2

Problems created
by poster boards
positioned
beside a wall.

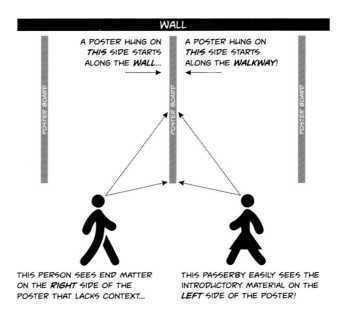

bottom poster you need to bring your eye level somewhere around the height of the presenter's crotch.

Never hold a poster session outdoors. This is just asking for the weather gods to laugh at you.

Lighting

The room for the poster session should be well lit. Particularly when conferences are held in hotels, the lighting in too many meeting rooms has been optimized for social events like wedding receptions, rather than viewing technical graphics and reading. Too many of these rooms are wannabe nightclubs. They're just too dim. And given that scientists tend to make dubious color choices, it can make it hard to read the poster. A venue that has windows and natural light is wonderful.

Water

All that talking can make you thirsty. Make sure there is a source of water readily and constantly available that doesn't require someone to leave the session for ten minutes. Other refreshments are nice, but nothing beats water.

Chairs

Chairs must be available for poster sessions. Many conferences generally do not have enough places for people to sit, but lack of seating is particularly a problem for poster sessions.

The expectation is that presenters will stand throughout a poster session, and this may be extremely difficult for some people. Not everyone is in the best of health when a conference happens. (Some time, I'll tell you about the student who got food poisoning while flying to a conference.) There are many reasons that people might want to sit, including migraine, pregnancy, chronic pain, anemia, anxiety, and many others (McLaughlin 2019, Schiff 2019). But even people in good health can benefit from having chairs available. At big conferences, you can be on your feet all day. There's a lot of walking from room to room, and poster sessions are generally several hours long. Even the healthiest and heartiest can start to flag a little under those conditions.

Bar stools would be even better. At regular sitting height, people cannot easily see the presenter. If someone does come over, a viewer may have to bend or crouch to be able to hear the presenter, particularly if the session is busy (MD 2012).

There are occasional reports of conference organizers or security who not only omit chairs, but refuse to provide them when asked (Emily 2012, McLaughlin 2019). This is not an acceptable way to treat conference attendees. Yes, having chairs in the poster hall does require enough space and a little planning. But some of your attendees will need chairs, and will thank you for providing them.

Venue checks

Organizers should schedule a site visit of the venue and include a walkthrough of the poster-session rooms. This is an opportunity to ensure that the poster session has enough amenities, like convenient access to bathrooms, water, and power outlets. It is also a chance to ensure that the lighting is sufficiently bright, which is difficult to tell from photographs.

Scheduling poster sessions

Many conference organizers do not allot enough time for poster viewing. While in theory a person might be able to see five posters in an hour – twelve minutes per poster sounds like a lot compared to most presentations – the reality is that unlike conference talks, poster sessions are unstructured and have more distractions. You might run into people who are not presenting posters. Presenters might walk away from their posters, so that audience members must come back. And there might be food and drinks, which can be more desirable than looking at a poster, especially at the end of a long day.

If possible, have posters go up in the room early, many hours before the allotted time for the poster session begins. This gives people a chance to browse, see what posters they most want to talk to the presenter about, and set their priorities for the poster session. It also provides a good "down time" for conference goers who want a quiet moment away from crowds, but still want to take in the academic content they are at the meeting to see.

Scheduling for poster sessions should ideally include some time when there are only poster sessions and no competing oral presentations. At very large meetings, where simultaneous sessions are just a fact of life, this may not be possible. A poster session has more simultaneous threads of programming than oral presentations, so a full slate of talks going on simultaneously with a poster session will cut attendance.

Schedule more than one poster session during the meeting. While presenters are often instructed to stay near their poster for part of the session, the reality is that there will often be a continual stream of people who want to talk to the presenter, and presenters may not be able to tear themselves away. Nor should they!

On a related note, split up posters on related topics into different sessions. It's tempting to put all the posters on a single related topic together. It makes them easier to find. But don't forget, people who are presenting posters are often the very people who most want to see those other posters on the same topic. Give presenters a chance to look around; they want to see other stuff, too.

Chapter recap

- Conference organizers should resist the temptation to try to fit more posters into a space. Small posters make for uncomfortable viewing and create accessibility problems.

- Poster halls should be well lit, near water and washrooms, and have chairs available.

- Getting to see posters that you want to see is a recurring problem, so poster sessions should be longer rather than shorter, and should not conflict with other programming tracks.

CHAPTER 27

Conference website resources

Presenter instructions

A conference website should have clear instructions for poster presenters. The most important piece of information should be the maximum size of the poster, because this is such a common problem. This should be the first thing listed, and it should be extremely large, with a picture showing whether the size is in portrait or landscape format.

Because not all poster designers can be assumed to have read this book, instructions provided by the organizers should inform poster presenters of several best practices in designing posters, such as designing the poster for accessibility.

Do not tell presenters to put abstracts on posters. The abstracts should be available in conference books, website, or app. There should be no need to put an abstract on the poster.

It's also unnecessary to ask presenters to put poster numbers on their titles. It should be the responsibility of the conference organizers to mark poster boards, not the presenters.

There should never be a template that is required for all presenters, or a request that people include conference logos. Everyone knows what conference they are attending.

Program book

There should be a way to search the abstracts and generate an itinerary.

It's useful to compile all the poster abstracts into a program book, even with a simple static format.

Because many abstracts never make it into the literature (Scherer *et al.* 2018), there should be an archive of all documents, preferably in a standard form like a PDF. Ideally, the document or even individual abstracts would be assigned a digital object identifier (DOI) to improve findability and stability (Foster *et al.* 2019).

Invitation services

New researchers often want to go to conferences specifically to meet particular people. But this can be hard, because it can be difficult to track a person down at a conference. The bigger the conference, the harder this gets.

Even if you can find someone, they might be deep in conversation with someone else and they might be intimidated. And some poster presenters are shy (academics are often a little introverted). It can be tough for a presenter, particularly someone who is new to the field, to search down people during a conference who might be interested and say, "Hey, please come to my poster." Conferences can facilitate these conversations by having poster invitation services. On the conference website, create a form and rules for inviting people to view your poster. Typically, there are limits to how many people can be invited – three is a good number. You may want to provide guidance on who can or should be invited. Email the invited viewer the name of the person inviting them, contact information, poster title, abstract, and viewing information (hall location, poster number, etc.)

Remind people that an invite is not a guarantee that the person will come to the poster.

Code of conduct

As noted in Chapter 24, poster presenters may sometimes be confronted by audience members behaving badly. Conference organizers should provide very clear codes of conduct on their website regarding appropriate behavior in conferences, including poster sessions (Foxx *et al.* 2019). Such codes should describe how conference attendees can report issues, and issues should be reported to impartial third parties. A code of conduct should also describe the consequences of violating the code (Foxx *et al.* 2019).

Social media policies

Organizers should clearly state their policies about whether they permit sharing posters via social media. Be aware that people expect to be able to take selfies in front of their own posters, and pictures of other people's posters for notetaking purposes, and that some posters will use QR codes for supplemental information. In most cases, a "no photos on the poster floor" policy is going to fail badly. Archive posters before the conference. "Small changes, like making presentations and posters available online before talks, are within the purview of societies and conference organizers, and could make an important difference to users of assistive technology such as screen readers" (Goring *et al.* 2018).

Chapter recap

- Conference websites are the first point of contact for poster presenters and need clear information for them.

- Conference organizers should provide a program book (not necessarily physically printed) that serves as a guide to content during the meeting and provides some archival value.

- Conference organizers should set clear expectations for the behavior of people attending.

CHAPTER 28

During the conference

Staying on schedule

Smaller conferences may have only a single track of oral presentations, with poster sessions scheduled around these. When faced with presenters going badly over time, too many organizers spontaneously cut back on poster viewing time rather than calling out the rude and unprofessional behavior of their keynote speakers. But it is deeply unfair to accommodate the poor preparation of a single person, who is often very senior, over the completed presentations of many people, who are often students. Organizers must have strong moderators to ensure that oral sessions finish on time.

Poster session staff

Conference organizers should ensure that there are clearly identifiable staff members who can assist poster presenters and viewers, particularly those who are low vision or blind. "Poster halls are often giant and difficult to navigate, so this is a great example of a disability-related accommodation that would benefit everyone" (Serrato Marks 2018). Staff members can also help provide some security in case of emergency, even if you don't have an actual bouncer.

Supplies

Presenters expect that the conference will supply them with tacks, pins, or whatever fasteners they need. In 2010, one of the largest academic conferences, the Society for Neuroscience, ordered 110,000 push pins for

15,116 poster presentations. That works out to over seven pins per poster, assuming no pins were reused (which they surely were).

Social events

Some conferences combine poster sessions with other events, such as a coffee cart, bar services, or other networking events. For example, some conferences turn their poster sessions into "happy hour" and give drink tickets to poster presenters to facilitate networking. The idea is that anyone who wants a drink must go visit a poster and talk to the presenter.

Conference attendees can have mixed feelings about the presence of alcohol in a professional meeting. One the one hand, many people enjoy alcohol responsibly and find it makes social encounters easier and more enjoyable. On the other hand, there are people who do not drink alcohol because they are underage, or abstain for personal, medical, or religious, reasons. Some people try to excuse bad behavior by saying, "I drank too much." But people can behave badly with or without alcohol.

Chapter recap

- Never take time away from poster sessions to fix delays in scheduling.
- Poster sessions benefit from staffing.
- There are pros and cons to having social events run concurrently with poster sessions.

PART IV

What next?

CHAPTER 29

Constant improvement

I have bad news. You bought this book, made it all the way to the end, and I am here to warn you that creating your posters could still be a frustrating experience and you may not always be happy with your results. Posters are not a solved problem, and you can't master a complex craft like graphic design in a single session. You must practice. Take pictures of things you like, that you think are well designed. Analyze what it is about them that you like.

You need to practice because the bar for poster presentations is going to be raised on you. Practically any visual medium becomes ever more sophisticated. Early movies look like stage plays. Late-nineteenth-century newspapers were little more than thin columns of text. Early comics were closer to text stories with illustration than graphic storytelling (Cohn *et al.* 2017).

There are two reasons for this progression.

The first reason is technical progress. Film cameras get smaller and more portable, color film is developed, and eventually it is all replaced by high-definition video. Newspapers move from being created using hot metal to being made on computers. Comics move from cheap newsprint and four colors to higher-quality paper and full color.

The second reason is creative innovation. Professionals working in the medium expanded the visual vocabulary available to themselves and other practitioners. Movie directors and cinematographers took years to realize that the rules of film are not the rules of theatre. Filmmakers had to invent techniques like "freeze frame," "dolly zoom," and "bullet time," which became part of film's language. Early comic artists mostly used rectangular grids with lots of descriptive text, but artists continually experimented with layouts, making them diverse and less reliant on text (Spiegelman 2002, McGovern 2014, Cohn *et al.* 2017), and comic letterers developed punctuation unique to the medium (Klein 2010).

These same two factors are at play in academic posters. On the technical side, posters have gone from being a collage of letter-sized (or A4) pages to full-size single sheets, with digital posters on screens looming large on the horizon.

What has held back conference posters as a medium (that is, why so many posters are bad) – and may continue to do so – is that academia has collective amnesia when it comes to posters.

First, there is a lack of continuity in the people who create posters. For example, two-thirds of the authors of conference abstracts only attended the Society for Neuroscience meeting once over six years (Lin *et al.* 2008). This is probably because many conference abstracts are authored by students and postdocs who do not stay in the field. It is a pattern that isn't easily changed. The constant turnover of poster makers means that many posters will always be made by beginners, slowing down the improvement of conference posters.

Second, there is almost no published body of work on conference poster design. It's no accident that so much of this book draws from web pages, blogs, and social media – there's not much else. In contrast, old advertising and propaganda posters are regularly collected in large coffee-table books. These posters were meant to be of the moment and almost disposable rather than high art. But they have been given serious attention and recognized on their own merits. Likewise, filmmakers and comic artists routinely discuss how works that are decades old inspired them (Spiegelman 2002). Without 1930s and 1940s movie serials, there would be no *Raiders of the Lost Ark*. Without *Flash Gordon*, there would be no *Star Wars*. Without *Creature from the Black Lagoon*, there would be no *Shape of Water*. Their sources of inspiration are often obscure and not all that widely known in popular culture, but they push the art forward. Conference poster creators have not been able to look back, see conference posters from past decades, and steal the sentiment. Poster makers might, if they are lucky, see a few posters from people in their department that are a few years old. Academia badly needs to start remembering its conference posters if it is to learn and improve.

But the prospects for improvement on this count are good. As more and more conferences archive posters, and as more people upload their own posters onto their own websites, blogs, and servers, a body of work will start to emerge. Hopefully, books like this can contribute by treating posters not as disposable, but as a form of scholarly communication that is unique, legitimate, and worth treating seriously as part of the academic literature.

Chapter recap

- Posters will never get better until there is an archived body of work that people can refer to and that practitioners can build on.

Afterword

This book is, in part, intended as an act of subversion.

In my time as an academic, I have recognized there is a sort of person who is actively hostile to clarity or claims not to care about design. (Weirdly, people who claim not to care about design will often object strenuously when someone creates a design that they do not like.)

Peter Jacobs (2019) wrote:

> It's important to understand that there is a subculture and in many places the dominant culture that responds negatively to such clarified presentations and deems them "not real science." … I have seen older and younger scientists alike react hostilely to these sorts of presentations – demanding to know where the equations are; where the error bars are; where the laboriously detailed methods description is.

Jacquelyn Gill (2019) replied:

> I have stories where the clarity of our presentation was interpreted as evidence that our research was too simple/not novel enough/ not rigorous, or not our own work (?!), because "I understood it all."

Jonathan Foley (2019) summed it up thus:

> Somehow there's a feeling in some corners of science that "under- standable" means less rigorous thinking, when in fact it's the opposite.

This book is an argument that even very focused training in science, technology, engineering, and mathematics (STEM) would benefit from including the arts, too. (I said that design isn't art at the start of this book, but design is taught alongside and with the arts more often than it is with the sciences.) I am not the first person to suggest that "STEM" become "STEAM," and I realize that it is impossible to teach someone every possible thing that might benefit them.

But scholarship only has value when it is shared. Sharing is, by definition, an experience that is social and embodied. There is no such thing as "pure" information. Information is always embedded in a physical representation, like a poster, which must be designed. Design – decision making guided by empathy – is a powerful aid to the sciences because it pushes back on the notion that "data" and "information" are the only things that matter. Considering other people – their experiences, their abilities, their preferences – matters. That's what I have learned from trying to make better posters.

References

Unless stated otherwise, all website addresses included in this list were checked and verified in February 2020.

Abrizah, A. (2017) The cautious faculty: their awareness and attitudes towards institutional repositories. *Malaysian Journal of Library and Information Science* 14 (2): 17–37.

Adams, C. (2001) What does the filler text "lorem ipsum" mean? www.straightdope.com/columns/read/2290/what-does-the-filler-text-lorem-ipsum-mean.

Adams, D. (1985) *The Hitchhiker's Trilogy*, Omnibus Edition. New York: Harmony.

Alley, M. (2018) *The Craft of Scientific Writing*. New York: Springer-Verlag.

Allocca, K. (2011) Why videos go viral. *TEDYouth*. www.ted.com/talks/kevin_allocca_why_videos_go_viral.

Alter, A.L., Oppenheimer, D.M., Epley, N., and Eyre, R.N. (2007) Overcoming intuition: metacognitive difficulty activates analytic reasoning. *Journal of Experimental Psychology: General* 136 (4): 569–576.

Anonymous (2012) In praise of posters. *Nature Chemistry* 4 (2): 67.

Anscombe, F.J. (1973) Graphs in statistical analysis. *The American Statistician* 27 (1): 17–21.

Antonelli, P. (2007) Design and the elastic mind. *TED.* www.ted.com/talks/paola_antonelli_design_and_the_elastic_mind.

Arditi, A. (2017) Rethinking ADA signage standards for low-vision accessibility. *Journal of Vision* 17 (5): 8.

Attai, D.J., Radford, D.M., and Cowher, M.S. (2016) Tweeting the meeting: Twitter use at the American Society of Breast Surgeons Annual Meeting 2013–2016. *Annals of Surgical Oncology* 23 (10): 3418–3422.

Babb-Biernacki, S. [@Babberwocky] (2019a) https://twitter.com/Babberwocky/status/1161633332381257730.

Babb-Biernacki, S. [@Babberwocky] (2019b) https://twitter.com/Babberwocky/status/1161263116463824898.

Bakeman, R. and Gottman, J. (1986) *Observing Interaction: An Introduction to Sequential Analysis*. Cambridge: Cambridge University Press.

Bang, M. (2000) *Picture This: How Pictures Work*. New York: SeaStar Books.

Banzai, B. (2002) Pinky Carruthers' Unknown Facts: subtitle trivia track DVD bonus feature, *The Adventures of Buckaroo Banzai Across the 8th Dimension*: Special Edition. MGM.

Barlow, J., Stephens, P.A., Bode, M., Cadotte, M.W., Lucas, K., Newton, E., Nuñez, M.A., and Pettorelli, N. (2018) On the extinction of the single-authored paper: the causes and consequences of increasingly collaborative applied ecological research. *Journal of Applied Ecology* 55 (1): 1–4.

Bastian, H. (2018) Building a great scientific abstract: a quick checklist. *Absolutely Maybe.* http://blogs.plos.org/absolutely-maybe/2018/06/06/building-a-great-scientific-abstract-a-quick-checklist.

Bastian, H. (2019) Why don't we do more visualizations of methods? *Absolutely Maybe.* https://blogs.plos.org/absolutely-maybe/2019/06/05/why-dont-we-do-more-visualizations-of-methods.

Becker, R. (2018) Stock photos of scientists reveal that science is mostly about staring. www.theverge.com/tldr/2018/5/3/17316640/science-stock-photos-representation-fails-bad-stock-photos-of-my-job.

Bennett, C. (2009) The story behind the Atlantic salmon. http://prefrontal.org/blog/2009/09/the-story-behind-the-atlantic-salmon.

Bennett, C.M., Baird, A.A., Miller, M.B., and Wolford, G.L. (2009) Neural correlates of interspecies perspective taking in the post-mortem Atlantic Salmon: an argument for multiple comparisons correction. *NeuroImage* 47: S147.

Berger, J. and Milkman, K.L. (2012) What makes online content viral? *Journal of Marketing Research* 49 (2): 192–205.

Berger, J. and Milkman, K.L. (2013) Emotion and virality: what makes online content go viral? *Insights* 5 (1): 18–23.

Bering, E. [@EaberingEdgar] (2019) https://twitter.com/EaberingEdgar/status/1150043342224408576.

Berkson, W. (2010) Part 2: Readability, affability, authority. *Reviving Caslon.* https://ilovetypography.com/2010/11/02/reviving-caslon-part-2-readability-affability-authority.

Bigman, A. (2012) The best and worst typefaces (and here's why). *99designs.* https://99designs.com/blog/tips/the-best-and-worst-typefaces-and-why.

Bik, H.M. and Goldstein, M.C. (2013) An introduction to social media for scientists. *PLoS Biology* 11 (4): e1001535.

Billings, S.B. and Schnepel, K.T. (2018) Life after lead: effects of early interventions for children exposed to lead. *American Economic Journal: Applied Economics* 10 (3): 315–344.

Blakley, J. (2010) Lessons from fashion's free culture. *TED.* www.ted.com/talks/johanna_blakley_lessons_from_fashion_s_free_culture.

Blankenship, J. and Dansereau, D.F. (2000) The effect of animated node-link displays on information recall. *Journal of Experimental Education* 68 (4): 293–308.

Boardley, J. (2008) On choosing type. *I Love Typography.* https://ilovetypography.com/2008/04/04/on-choosing-type.

Bombaci, S.P., Farr, C.M., Gallo, H.T., Mangan, A.M., Stinson, L.T., Kaushik, M., and Pejchar, L. (2016) Using Twitter to communicate conservation science from a professional conference. *Conservation Biology* 30 (1): 216–225.

Borrego, Á. (2017) Institutional repositories versus ResearchGate: the depositing habits of Spanish researchers. *Learned Publishing* 30 (3): 185–192.

Botha, E. and Reyneke, M. (2013) To share or not to share: the role of content and emotion in viral marketing. *Journal of Public Affairs* 13 (2): 160–171.

Bricker, D. (2013) How many spaces after a period? Ending the debate. http://theworldsgreatestbook.com/how-many-spaces-after-a-period.

Briscoe, M.H. (1996) *Preparing Scientific Illustrations*. New York: Springer-Verlag.

Brookshire, B. [@BeeBrookshire] (2019) https://twitter.com/BeeBrookshire/status/1143541861266518016.

Brumberger, E.R. (2003) The rhetoric of typography: the persona of typeface and text. *Technical Communication* 50 (2): 206–223.

Bumiller, E. (2010) We have met the enemy and he is PowerPoint. *New York Times*. www.nytimes.com/2010/04/27/world/27powerpoint.html.

Burnham, T. (2015) A trick for higher SAT scores? Unfortunately no. *Observations of a Biological Economist*. www.terryburnham.com/2015/04/a-trick-for-higher-sat-scores.html.

Butler, J.R. (2015) Proof that using Creative Commons material is not risk-free. *Guide Through the Legal Jungle*. www.lexology.com/library/detail.aspx?g=1a0770bf-ae7f-4206-a352-ca5d5657414d.

Butterick, M. (2016) *Butterick's Practical Typography*. https://practicaltypography.com.

Campbell, J. (2010) Sunday surgery: (not quite) new year's resolution. *Jim Cambell, Man of Letters*. http://clintflickerlettering.blogspot.com/2010/11/sunday-surgery-not-quite-new-years.html.

Cárcamo Ulloa, L., Marcos Mora, M.-C., Cladellas Pros, R., and Castelló Tarrida, A. (2015) News photography for Facebook: effects of images on the visual behaviour of readers in three simulated newspaper formats. *Information Research* 20 (1): 660.

Carroll, L. (1865) *Alice's Adventures in Wonderland*. London: Macmillan.

Carson, J. (2018) Why your headlines are worth almost all your content marketing efforts (and how to improve them). *Econsultancy*. https://econsultancy.com/why-your-headlines-are-worth-almost-all-your-content-marketing-efforts-and-how-to-improve-them.

Carswell, C.M., Frankenberger, S., and Bernhard, D. (1991) Graphing in depth: perspectives on the use of three-dimensional graphs to represent lower-dimensional data. *Behaviour & Information Technology* 10 (6): 459–474.

Castelvecchi, D. (2015) Physics paper sets record with more than 5,000 authors. *Nature*. doi: 10.1038/nature.2015.17567.

Cetina, K.K. (1997) Sociality with objects: social relations in postsocial knowledge societies. *Theory, Culture and Society* 14 (4): 1–30.

Chawla, D.S. (2017) Publishers take academic networking site to court. *Science* 358 (6360): 161.

Cheplygina, V. (2017) How to recycle your fabric poster. *Dr Veronika Ch*. https://veronikach.com/lifestyle/how-to-recycle-your-fabric-poster-faq.

Cleveland, W.S. and McGill, R. (1985) Graphical perception and graphical methods for analyzing scientific data. *Science* 229 (4716): 828–833.

Cohn, N. (2013) Navigating comics: an empirical and theoretical approach to strategies of reading comic page layouts. *Frontiers in Psychology* 4: 186.

Cohn, N. [@visual_linguist] (2017) https://twitter.com/visual_linguist/status/877985271060598784.

Cohn, N. and Campbell, H. (2015) Navigating comics II: Constraints on the reading order of comic page layouts. *Applied Cognitive Psychology* 29 (2): 193–199.

Cohn, N., Taylor, R., and Pederson, K. (2017) A picture is worth more words over time: multimodality and narrative structure across eight decades of American superhero comics. *Multimodal Communication* 6 (1): 19–38.

Collinge, R. (2017) How to design for dyslexia. *Usabilla Blog*. http://blog.usabilla. com/how-to-design-for-dyslexia.

Columbia Accident Investigation Board (2003) *Columbia Accident Investigation Board Report Volume 1, NASA*. 1: 248.

CONSORT (2010) CONSORT 2010 flow diagram. www.consort-statement.org/ consort-statement/flow-diagram.

Cook, A.R. and Teo, S.W.L. (2011) The communicability of graphical alternatives to tabular displays of statistical simulation studies. *PLoS ONE* 6 (11): e27974.

Creative Commons (2019) CC licenses and examples. https://creativecommons. org/share-your-work/licensing-examples.

Creed, M. [@Meaghan_Creed] (2018) https://twitter.com/Meaghan_Creed/status/ 1027050192007319552.

Crow, R. (2002) *The Case for Institutional Repositories: A SPARC Position Paper*. Washington, D.C.: The Scholarly Publishing and Academic Resources Coalition.

Crowe, D. (2019) Visual and UX design principles can improve the effectiveness of poster sessions. https://derekcrowe.net/butterposter.

Cumming, G., Fidler, F., and Vaux, D.L. (2007) Error bars in experimental biology. *Journal of Cell Biology* 177 (1): 7–11.

Dabner, D., Calvert, S., and Casey, A. (2010) *Graphic Design School: The Principles and Practices of Graphic Design*. London: QuatroPublishing.

Dayas, C. [@neuronewy] (2018). https://twitter.com/neuronewy/status/1059126 228399976448.

de Vries, S. (2017) REpost Science. www.repost-science.nl.

del Toro, G. (2013) *Pacific Rim*. Warner Bros., 131 minutes.

Department of Justice (2010) 2010 ADA Standards for Accessible Design. www. ada.gov/2010ADAstandards_index.htm.

Devitz, A.-C. (2019) SICB 2019 – an accessibility perspective. *The Bendy Biologist*. https://bendybiologist.com/2019/01/13/sicb-2019-an-accessibility -perspective.

Di Girolamo, N. and Reynders, R.M. (2017) Health care articles with simple and declarative titles were more likely to be in the Altmetric Top 100. *Journal of Clinical Epidemiology* 85: 32–36.

Diemand-Yauman, C., Oppenheimer, D.M., and Vaughan, E.B. (2011) Fortune favors the bold (and the italicized): effects of disfluency on educational outcomes. *Cognition* 118 (1): 111–115.

Djamasbi, S., Siegel, M., and Tullis, T. (2010) Generation Y, web design, and eye tracking. *International Journal of Human-Computer Studies* 68 (5): 307–323.

Douglas, N. (2016) Ridiculous stock photo clichés we can probably retire now. *The Daily Dot*. www.dailydot.com/unclick/most-ridiculous-news-stock -photo-cliches.

Drake, M.A. (2004) Institutional repositories: hidden treasures. *Searcher* 12 (5).

Drauglis v. Kappa Map Group, LLC. (2015) United States District Court for the District of Columbia. https://dockets.justia.com/docket/district-of -columbia/dcdce/1:2014cv01043/166825.

Duarte, N. (2009a) *Slide:ology: The Art and Science of Creating Great Presentations*. Sebastopol, California: O'Reilly Media.

Duarte, N. (2009b) Stanford passed (and failed) the Glance Test. Would you? www.duarte.com/presentation-skills-resources/stanford-passed-and-failed-the-glance-test-would-you.

Duarte, N. (2019a) *Data Story: Explain Data and Inspire Action Through Story*. Oakton, Virginia: Ideapress Publishing.

Duarte, N. (2019b) Three ways to effectively communicate to different kinds of decision-makers. https://thriveglobal.com/stories/three-ways-to-effectively-communicate-to-different-kinds-of-decision-makers/?utm_campaign=coschedule&utm_source=linkedin&utm_medium=Nancy%20Duarte.

Duffy, M.A. (2017) Last and corresponding authorship practices in ecology. *Ecology and Evolution* 7 (21): 8876–8887.

Editorial Team (2017) The terrible 20 fonts you should absolutely avoid using. https://1stwebdesigner.com/bad-fonts.

Eisen, J. (2011a) More art and science – hand painted poster at Synthetic Biology #synbio5 – by Karmella Haynes. *The Tree of Life*. https://phylogenomics.blogspot.com/2011/06/more-art-science-hand-painted-poster-at.html.

Eisen, J. (2011b) More pics of hand painted poster from #synbio5. *The Tree of Life*. https://phylogenomics.blogspot.com/2011/06/more-pics-of-hand-painted-poster-from.html.

Ekins, S. and Perlstein, E.O. (2014) Ten simple rules of live tweeting at scientific conferences. *PLoS Computational Biology* 10 (8): e1003789.

Elam, K. (2004) *Grid Systems: Principles of Organizing Type*. Hudson, New York: Princeton Architectural Press.

Else, H. (2018) Major publishers sue ResearchGate over copyright infringement. *Nature*. doi: 10.1038/d41586-018-06945-6.

Emily (2012) Sick at conferences (comment). *Gravity's Rainbow*. http://sarcozona.org/2012/10/15/sick-at-conferences/#comment-11265.

Eng, K.F. (2012) How to beat the TED time limit: fellows give 20-second TEDTalks. *TEDBlog*. https://blog.ted.com/how-to-beat-the-ted-time-limit-fellows-give-20-second-tedtalks.

Engeström, J. (2005) Why some social network services work and others don't – Or: the case for object-centered sociality. *Zengestrom*. www.zengestrom.com/blog/2005/04/why-some-social-network-services-work-and-others-dont-or-the-case-for-object-centered-sociality.html.

Faraday, P. (2000) Visually critiquing web pages. 6th Conference on Human Factors and the Web, Austin, Texas.

Faulkes, Z. (2007) Motor neurons involved in escape responses in white shrimp, *Litopeneaus* [sic] *setiferus*. *Integrative and Comparative Biology* 47 (Supplement 1): e178.

Faulkes, Z. (2010) Don't dangle. *Better Posters*. http://betterposters.blogspot.com/2010/11/dont-dangle.html.

Faulkes, Z. [@DoctorZen] (2019) https://twitter.com/DoctorZen/status/1141393878630772737.

Faulkes, Z. and Varghese, N. (2004) No escape: loss of escape-related giant neurons in spiny lobsters (*Palinurus* (sic) *argus*). Society for Neuroscience 2004. San Diego, California, USA, Society for Neuroscience.

Few, S. (2007) Save the pies for dessert. Perceptual Edge. www.perceptualedge.com/articles/visual_business_intelligence/save_the_pies_for_dessert.pdf.

Few, S. (2009) *Now You See It*. Oakland, California: Analytics Press.

Fisher, M. and Keil, F.C. (2016) The curse of expertise: when more knowledge leads to miscalibrated explanatory insight. *Cognitive Science* 40 (5): 1251–1269.

Fleerackers, A. (2019) Emojis in scholarly communication: [lit] or [shit]? *scholcommlab.* www.scholcommlab.ca/2019/08/21/emojis-in-scholcomm.

Flesch, R. (1946) *The Art of Plain Talk.* New York: Harper & Row.

Foley, J. [@GlobalEcoGuy] (2019) https://twitter.com/GlobalEcoGuy/status/1136606582635212800.

Foster, C., Wager, E., Marchington, J., Patel, M., Banner, S., Kennard, N.C., Panayi, A., Stacey, R., and the GPCAP Working Group. (2019) Good practice for conference abstracts and presentations: GPCAP. *Research Integrity and Peer Review* 4 (1): 11.

Fournier, A. [@RallidaeRule] (2019) https://twitter.com/RallidaeRule/status/1176837104602664960.

Fox, C.W. and Burns, C.S. (2015) The relationship between manuscript title structure and success: editorial decisions and citation performance for an ecological journal. *Ecology and Evolution* 5 (10): 1970–1980.

Foxx, A.J., Barak, R.S., Lichtenberger, T.M., Richardson, L.K., Rodgers, A.J., and Webb Williams, E. (2019) Evaluating the prevalence and quality of conference codes of conduct. *Proceedings of the National Academy of Sciences* 116 (30): 14931–14936.

Freitag, A. (2012) How to make a pretty map. *Southern Fried Science.* www.southernfriedscience.com/how-to-make-a-pretty-map.

Fry, B. (2019) The four Cs of data + design. https://medium.com/@ben_fry/the-four-cs-of-data-design-6eadd69b05da.

Garfield, S. (2011) The 8 worst fonts in the world. *Fast Company.* www.fastcompany.com/1665318/the-8-worst-fonts-in-the-world.

Gelman, A., Pasarica, C., and Dodhia, R. (2002) Let's practice what we preach: turning tables into graphs. *The American Statistician* 56 (2): 121–130.

Gendelman, V. (2013) Worst Fonts Ever! 11 Examples of Bad Typography in Print. *Company Folders.* www.companyfolders.com/blog/worst-fonts-ever-11-examples-of-bad-typography-in-print.

Gianotti, F. (2012) Status of standard model Higgs searches in ATLAS. https://indico.cern.ch/getFile.py/access?contribId=1&resId=1&materialId=slides&confId=197461.

Gill, J. [@JacquelynGill] (2019) https://twitter.com/JacquelynGill/status/1136602011401629697.

GISGeography (2020) The ultimate list of GIS formats and geospatial file extensions. https://gisgeography.com/gis-formats.

Gladwell, M. (2013) *David and Goliath: Underdogs, Misfits, and the Art of Battling Giants.* New York: Little, Brown and Company.

Gloviczki, P. and Lawrence, P.F. (2018) Visual abstracts bring key message of scientific research. *Journal of Vascular Surgery* 67 (5): 1319–1320.

Godin, S. (2003a) How to get your ideas to spread. *TED.*

Godin, S. (2003b) Remarkable is where you choose to make it. *Seth's Blog.* https://seths.blog/2003/06/remarkable_is_w.

Godin, S. (2010) Type tells a story. *Seth's Blog.* https://seths.blog/2010/01/type-tells-a-story.

Gopen, G.D. and Swan, J.A. (1990) The science of scientific writing. *American Scientist* 78 (November–December): 550–558.

Goring, S.J., Whitney, K.S., and Jacob, A.L. (2018) Accessibility is imperative for inclusion. *Frontiers in Ecology and the Environment* 16 (2): 63.

Granados, M. (2018) Museum inspired scientific posters. *Descienceblog*. https://descienceblog.tumblr.com/post/176490405665/museum-inspired-scientific-posters-the-2018-north.

Greenfieldboyce, N. (2019) To save the science poster, researchers want to kill it and start over. www.npr.org/sections/health-shots/2019/06/11/729314248/to-save-the-science-poster-researchers-want-to-kill-it-and-start-over.

Gregory, M.W. (1992) The infectiousness of pompous prose. *Nature* 360 (6399): 11–12.

Gurwell, M. (2012) Hang time. *Better Posters*. http://betterposters.blogspot.com/2012/11/hang-time.html.

Hampton-Smith, S. (2017) The 8 biggest typography mistakes designers make. *Creative Bloq*. www.creativebloq.com/typography/mistakes-41411451.

Hansen, J. (2010) So you need a typeface. https://julianhansen.com/soyouneedatypeface.

Harpers Ferry Center Accessibility Committee. (2017) Programmatic accessibility guidelines for National Park Service interpretive media. www.nps.gov/features/hfc/guidelines.

Harrison, K.H. [@khhsocratica] (2018) https://twitter.com/khhsocratica/status/1026952981579628544.

Hartley, J. and Sheridan, V. (2019) Resources for writing academic English when English is not your first language. https://ease.org.uk/publications/ease-toolkit-authors/resources-for-writing-academic-english-when-english-is-not-your-first-language.

Heraclitus (2011) Why two spaces after a period isn't wrong (or, the lies typographers tell about history). *Heraclitean River*. https://web.archive.org/web/20171217060354/http://www.heracliteanriver.com/?p=324.

Hernandez, D. [@deray28] (2017) https://twitter.com/deray28/status/883812844348768256.

Hess, G., Tosney, K., and Liegel, L. (2013) Creating effective poster presentations | an effective poster. https://projects.ncsu.edu/project/posters.

Hinds, P.J. (1999) The curse of expertise: The effects of expertise and debiasing methods on prediction of novice performance. *Journal of Experimental Psychology: Applied* 5 (2): 205–221.

Ho, J., Tumkaya, T., Aryal, S., Choi, H., and Claridge-Chang, A. (2019) Moving beyond P values: data analysis with estimation graphics. *Nature Methods* 16 (7): 565–566.

Hockin, B. [@bryonyhockin] (8 August 2018) https://twitter.com/bryonyhockin/status/1027209616449576960.

Holmberg, N. (2004) Eye movement patterns and newspaper design factors: an experimental approach. Master's thesis, Lund University.

Holsanova, J., Holmberg, N., and Holmqvist, K. (2005) Tracing integration of text and pictures in newspaper reading. *Lund University Cognitive Studies* 125: 1–19.

Houston, D. (1979) Buck Rogers becomes the movie. *Starlog* 4 (21): 50–52. https://archive.org/details/starlog_magazine-021.

Huff, D. (1954) *How to Lie with Statistics*. New York: W.W. Norton.

Hustwit, G. (2007) *Helvetica*. Petit Grand Publishing: 80 minutes.

Hut, R. [@RolfHut] (2017) https://twitter.com/RolfHut/status/ 8546280742
43182593.

Ibrahim, A.M. (2018) Seeing is believing: Using visual abstracts to disseminate
scientific research. *American Journal of Gastroenterology* 113 (4): 459–461.

Ibrahim, A.M., Lillemoe, K.D., Klingensmith, M.E., and Dimick, J.B. (2017)
Visual abstracts to disseminate research on social media: a prospective, case–
control crossover study. *Annals of Surgery* 266 (6): e46–e48.

International Society for Traumatic Stress Studies (2019) Annual Meeting
Presenter Information. www.istss.org/am19/annual-meeting-presenter
-information.aspx.

Jacobs, P. [@past_is_future] (2019) https://twitter.com/past_is_future/
status/1136600588421140480.

Jacques, T.S. and Sebire, N.J. (2010) The impact of article titles on citation hits:
an analysis of general and specialist medical journals. *JRSM Short Reports*
1 (1): 1–5.

Jamali, H.R. and Nikzad, M. (2011) Article title type and its relation with the
number of downloads and citations. *Scientometrics* 88 (2): 653–661.

Jambor, H. (2019) A quick guide to better figures. *helena * jambor*. https://
helenajambor.wordpress.com/2019/07/16/a-quick-guide-to-better-figures.

Jury, D. (2002) *About Face: Reviving the Rules of Typography*. Mies, Switzerland:
RotoVision.

Jury, D. (2006) *What is Typography?* Mies, Switzerland: RotoVision.

Kammeyer, J. (2009) Creating PowerPoint based on research. *Comm Comm*.
http://jenniferkammeyer.typepad.com/commcomm/2009/01/creating
-powerpoint-based-on-research.html.

Keegan, D.A. and Bannister, S.L. (2003) Effect of colour coordination of attire
with poster presentation on poster popularity. *CMAJ* 169 (12): 1291–1292.

Keith, J. (2007) www.flickr.com/photos/adactio/537789598.

Khoury, C.K., Kisel, Y., Kantar, M., Barber, E., Ricciardi, V., Klirs, C., Kucera, L.,
Mehrabi, Z., Johnson, N., Klabin, S., Valiño, Á., Nowakowski, K., Bartomeus,
I., Ramankutty, N., Miller, A., Schipanski, M., Gore, M.A., and Novy, A.
(2019) Science–graphic art partnerships to increase research impact.
Communications Biology 2 (1): 295.

Kindel, E., Walker, S., Burke, C., Eve, M., Minns, E., and Perks, S. (2009–2020)
Isotype Revisited. http://isotyperevisited.org.

Kinsky, E.S. and Bruce, K. (2016) "It throws you into the ring": learning from
live-tweeting. *Teaching Journalism and Mass Communication* 16 (1): 36–52.

KK (2012) How to dress for a conference like a fashionable lady scientist. *My
Laser Boyfriend*. http://laserboyfriend.blogspot.com/2012/10/how-to-dress
-for-conference-like.html.

Klein, T. (2009) Comic Sans font examined. *Todd's Blog*. https://kleinletters.com/
Blog/comic-sans-font-examined.

Klein, T. (2010) Punctuating comics: breath marks. *Todd's Blog*. https://
kleinletters.com/Blog/punctuating-comics-breath-marks.

Knowles, S. [@dr_know] (2019). https://twitter.com/dr_know/status/114639992
9138630656.

Kolko, J. (2018) We are illiterate. www.themoderniststudio.com/2018/08/20/we-are-illiterate.

Krause, K. (2019) The design decisions behind *Nature's* new look. *Nature* 574: 476–477.

Krino, G. (1989) Colors of the future. *Chicago Tribune*. http://articles.chicagotribune.com/1989-10-15/news/8901230024_1_color-forecasting-color-research-carlton-wagner.

Krzywinski, M. [@MKrzywinski] (2019) https://twitter.com/MKrzywinski/status/1128834516523479040.

Kühl, T. and Eitel, A. (2016) Effects of disfluency on cognitive and metacognitive processes and outcomes. *Metacognition and Learning* 11 (1): 1–13.

Kuruwita, R. [@reggiekuruwita] (2019) https://twitter.com/reggiekuruwita/status/1152114123959758853.

Landa, R., Gonnella, R., and Brower, S. (2007) *2D: Visual Basics for Designers.* Clifton Park, New York: Thomson/Delmar Learning.

Lang, T. (2017) How to shorten a text by up to 30% and improve clarity without losing information. *Medical Writing* 26 (1): 21–25.

Lanks, B. (2017) The father of grunge typography calls out lazy design. https://magenta.as/the-father-of-grunge-typography-calls-out-lazy-design-daae470a685a.

Lantz, C.L. (2018) Got a fabric poster? Sew yourself a tote! https://medium.com/@cllantz/got-a-fabric-poster-sew-yourself-a-tote-1b4cf60530dd.

Lin, J.M., Bohland, J.W., Andrews, P., Burns, G.A.P.C., Allen, C.B., and Mitra, P.P. (2008) An analysis of the abstracts presented at the annual meetings of the Society for Neuroscience from 2001 to 2006. *PLoS ONE* 3 (4): e2052.

Lockwood, G. (2016) Academic clickbait: articles with positively-framed titles, interesting phrasing, and no wordplay get more attention online. *The Winnower* 7: e146723.36330. doi: 10.15200/winn.146723.36330.

Loranger, H. (2015) Headings are pick-up lines: 5 tips for writing headlines that convert. *NN/g Nielsen Norman Group Articles.* www.nngroup.com/articles/headings-pickup-lines.

Loranger, H. and Nielsen, J. (2017) Microcontent: a few small words have a mega impact on business. *NN/g Nielsen Norman Group Articles.* www.nngroup.com/articles/microcontent-how-to-write-headlines-page-titles-and-subject-lines.

Lucas, F. (2019) Challenging the safety of conformity: better poster design to disseminate scientific knowledge fast. *YoungEHA.* https://ehaweb.org/youngeha/spotlight-on/challenging-the-safety-of-conformity-better-poster-design-to-disseminate-scientific-knowledge-fast.

Lupton, E. (2004) *Thinking With Type.* New York: Princeton Architectural Press.

Lupton, E. (2015) *How Posters Work.* New York: Cooper Hewitt, Smithsonian Design Museum.

Macgregor, J. (2018) The history of hit points. www.pcgamer.com/the-history-of-hit-points.

Mackiewicz, J. (2007) Audience perceptions of fonts in projected PowerPoint text slides. *Technical Communication* 54 (3): 295–307.

MacNamee, S. [@MacnameeSarah] (2019) https://twitter.com/MacnameeSarah/status/1027068104797564928.

Mandy, W. [@WillClinPsy] (2015) https://twitter.com/WillClinPsy/status/628932244837679104.

Mann, T. (1993) *Library Research Models: A Guide to Classification, Cataloging, and Computers*. Oxford: Oxford University Press.

Márquez, M.C. (2019) Say hello to a new relative of sharks. *Forbes*. www.forbes.com/sites/melissacristinamarquez/2019/09/09/say-hello-to-a-new-relative-of-sharks/#2c63ea24269c.

Matejka, J. and Fitzmaurice, G. (2017) Same stats, different graphs: generating datasets with varied appearance and identical statistics through simulated annealing. ACM SIGCHI Conference on Human Factors in Computing Systems, Denver, Colorado, SIGCHI.

McAndrew, F.T. and Koehnke, S.S. (2016) On the nature of creepiness. *New Ideas in Psychology* 43: 10–15.

McCloud, S. (1993) *Understanding Comics: The Invisible Art*. New York: William Morrow Paperbacks.

McCloud, S. (2018) The Big Triangle. www.scottmccloud.com/4-inventions/triangle.

McGovern, A. (2014) The frontier spirit: Will Eisner and the discovery of comics' next dimension. https://therumpus.net/2014/03/the-frontier-spirit-will-eisner-and-the-discovery-of-comics-next-dimension.

McGranahan, C. (2013) Conference chic, or, how to dress like an anthropologist. *Savage Minds*. https://savageminds.org/2013/11/20/conference-chic-or-how-to-dress-like-an-anthropologist.

McLaughlin, B.A. [@McLNeuro] (2018a) https://twitter.com/McLNeuro/status/1027242069125996545.

McLaughlin, B.A. [@McLNeuro] (2018b) https://twitter.com/McLNeuro/status/1059186046850592768.

McLaughlin, B.A. [@McLNeuro] (2019) https://twitter.com/McLNeuro/status/1145526185948012545.

McNutt, M. (2015) It starts with a poster. *Science* 347 (6226): 1047.

McSweeney, K. (2019) What does a scientist look like? *Diversity in Research Jobs*. www.diversityinresearch.careers/article/what-does-a-scientist-look-like-/.

MD (2012) Care to sit down? (comment). *Better Posters*. http://betterposters.blogspot.com/2012/11/care-to-sit-down.html?showComment=1351812391046#c7174585247441141964.

Meeks, E. (2018) You can design a good chart in R. *Towards Data Science*. https://towardsdatascience.com/you-can-design-a-good-chart-with-r-5d00ed7dd18e.

Mellow, G. (2018a) So you want to hire a science illustrator. *Symbiartic*. www.symbiartic.com/home/2018/11/28/so-you-want-to-hire-a-science-illustrator.

Mellow, G. (2018b) What if all the images went away. *Symbiartic*. www.symbiartic.com/home/scicomm-what-if-all-the-images-went-away.

Meyer, A., Frederick, S., Burnham, T.C., Guevara Pinto, J.D., Boyer, T.W., Ball, L.J., Pennycook, G., Ackerman, R., Thompson, V.A., and Schuldt, J.P. (2015) Disfluent fonts don't help people solve math problems. *Journal of Experimental Psychology: General* 144 (2): e16–e30.

Mock, L. (2018) The comprehensive guide to flowcharts. www.gliffy.com/blog/the-comprehensive-guide-to-flowcharts.

Moody, G. (2014) German court says Creative Commons "Non-commercial" licenses must be purely for personal use. *Techdirt*. www.techdirt.com/

articles/20140326/11405526695/german-court-says-creative-commons-non
-commercial-licenses-must-be-purely-personal-use.shtml.

Morris, A.B. (2017) Designing for dyslexia. *UX Planet.* https://uxplanet.org/
designing-for-dyslexia-6d12e8c41cd7.

Morrison, M. (2019) How to create a better research poster in less time (including
templates). *YouTube.* www.youtube.com/watch?v=1RwJbhkCA58.

Munroe, R. (2010) Convincing. https://xkcd.com/833.

Munroe, R. (2012) Ten thousand. https://xkcd.com/1053.

Murphy, S.M., Vidal, M.C., Hallagan, C.J., Broder, E.D., Barnes, E.E., Horna Lowell,
E.S., and Wilson, J.D. (2019) Does this title bug (Hemiptera) you? How to write
a title that increases your citations. *Ecological Entomology* 44 (5): 593–600.

Murray, E. (2019) The importance of color in data visualizations. *Forbes.* www.
forbes.com/sites/evamurray/2019/03/22/the-importance-of-color-in-data
-visualizations.

National Center for PTSD (2019) PTSD Checklist for DSM-5 (PCL-5). www.
ptsd.va.gov/professional/assessment/adult-sr/ptsd-checklist.asp.

O'Meara, B., Case, A., Wiggins, J., Zamudio, K., Baucom, R., and Strauss, S.
(2019) Unwelcome behaviors at the Evolution meetings: survey results.
Evolution, Providence, Rhode Island.

Ogilvy, D. (1963) *Confessions of an Advertising Man.* London: Southbank
Publishing.

Olson, R. (2014) Is Al Gore the worst climate communicator ever? *The Benshi.*
http://thebenshi.com/?p=5186.

Olson, R. (2015) *Houston, We Have a Narrative: Why Science Needs Story.*
Chicago, Illinois: University of Chicago Press.

Olson, R. [@ABTAgenda] (2016) https://twitter.com/ABTagenda/status/73077162
6338308097.

Olson, R. (2019a) *Narrative Is Everything: The ABT Framework and Narrative
Evolution.* Los Angeles, California: Prairie Starfish Productions.

Olson, R. [@ABTAgenda] (2019b) https://twitter.com/ABTagenda/status/
1166734801891545088.

Oppenheimer, D.M. (2006) Consequences of erudite vernacular utilized
irrespective of necessity: problems with using long words needlessly. *Applied
Cognitive Psychology* 20 (2): 139–156.

Orwell, G. (1946) Politics and the English language. *Horizon: A Review of
Literature and Art.* London. April.

Outing, S. (2004) Eyetrack III: What news websites look like through readers'
eyes. www.poynter.org/archive/2004/eyetrack-iii-what-news-websites-look
-like-through-readers-eyes.

Paiva, C.E., Lima, J.P.d.S.N., and Paiva, B.S.R. (2012) Articles with short titles
describing the results are cited more often. *Clinics* 67: 509–513.

Pal, P. (2020) 20 common typography mistakes that designers should not make in
2020. https://think360studio.com/blog/20-common-typography-mistakes
-that-designers-should-not-make-in-2020.

Parker, P., Chao, D., Norman, I., and Dunham, M. (2003) Orbiter assessment
of STS-107 ET bipod insulation ramp impact. [PowerPoint presentation.]
NASA.

Parks Canada (1993) Design guidelines for media accessibility. Ottawa:
Environment Canada.

Peck, E.M., Ayuso, S.E., and El-Etr, O. (2019) Data is personal: attitudes and perceptions of data visualization in rural Pennsylvania. Proceedings of the 2019 CHI Conference on Human Factors in Computing Systems. Glasgow, Scotland, ACM: 1–12.

Pederson, K. and Cohn, N. (2016) The changing pages of comics: page layouts across eight decades of American superhero comics. *Studies in Comics* 7 (1): 7–28.

Pemmaraju, N., Mesa, R.A., Majhail, N.S., and Thompson, M.A. (2017) The use and impact of Twitter at medical conferences: Best practices and Twitter etiquette. *Seminars in Hematology* 54 (4): 184–188.

Pinantoan, A. (2015) How to massively boost your blog traffic with these 5 awesome image stats. *BuzzSumo*. https://buzzsumo.com/blog/how-to -massively-boost-your-blog-traffic-with-these-5-awesome-image-stats.

Platt, J.R. (1964) Strong inference. *Science* 146 (3642): 347–353.

Purrington, C. (2019) Designing conference posters. http://colinpurrington. com/tips/poster-design.

Remes, K., Ortega, F., Fierro, I., Joger, U., Kosma, R., Marín Ferrer, J.M., Project PALDES, Niger Project SNHM, Ide, O.A., and Maga, A. (2009) A new basal sauropod dinosaur from the Middle Jurassic of Niger and the early evolution of Sauropoda. *PLoS ONE* 4 (9): e6924.

Renee, H. (2018) What to wear to an academic conference. *Shenova Fashion*. https://shenovafashion.com/blogs/blog/what-to-wear-to-an-academic -conference.

Reynolds, G. (2007) Who says we need our logo on every slide? *Presentation Zen*. www.presentationzen.com/presentationzen/2007/05/the_source_of_a.html.

Reynolds, G. (2011) *Presentation Zen: Simple Ideas on Presentation Design and Delivery*. Berkeley, California: New Riders.

Roam, D. (2013) *The Back of the Napkin: Solving Problems and Selling Ideas with Pictures*. New York: Portfolio.

Roberson, M. (2018) Finding myself in research. *New York Times*. www.nytimes. com/2018/05/23/opinion/graduate-student-cancer-research.html.

Roberts-Artal, L. (2016) GeoTalk: REcycle textile posters into useful products. *EGU Blogs*. https://blogs.egu.eu/geolog/2016/08/19/geotalk-recycle-textile -posters-into-useful-products.

Rodriguez, L. [@rodriguez_lion] (2018). https://twitter.com/rodriguez_lion/ status/1059159146396758016.

Rogers, S. (2014) What fuels a Tweet's engagement? *Media Blog*. https://blog. twitter.com/official/en_us/a/2014/what-fuels-a-tweets-engagement.html.

Rohde, J. [rockyrohde] (2019) www.instagram.com/p/B16se8eBhHY.

Rowe, N.E. (2014) Poster presentations – the "then and now" of a popular medium of scientific communication. *FEBS News* 2014: 9–10.

Rowe, N.E. and Ilic, D. (2015) Rethinking poster presentations at large-scale scientific meetings: is it time for the format to evolve? *The FEBS Journal* 282: 3661–3668.

Rutledge, A. (2008) On creativity. *A List Apart*. https://alistapart.com/article/ oncreativity.

Rutledge, A. (2009) Creativity is not design. *Design View*. Archived at https://web. archive.org/web/20091103155601/http://www.andyrutledge.com/creativity -is-not-design-test-2.php.

Rutledge, K. [@fishandfreckles] (2019) https://twitter.com/fishandfreckles/status/1167482593408770048.

Rutter, R. (2017) Web typography: designing tables to be read, not looked at. *A List Apart*. http://alistapart.com/article/web-typography-tables.

Sandquist, E. [@Ilovebraaains] (2014) https://twitter.com/Ilovebraaains/status/541289268207816704.

Sauro, J. (2016) Should all graphs start at 0? *Measuring U*. https://measuringu.com/graph-zero.

Scherer, R.W., Meerpohl, J.J., Pfeifer, N., Schmucker, C., Schwarzer, G., and von Elm, E. (2018) Full publication of results initially presented in abstracts. *Cochrane Database of Systematic Reviews* (11): MR000005.

Schiff, H.C. [@HillSchiff] (2019) https://twitter.com/HillSchiff/status/1145786007519997957.

Schneck, M.E., Haegerstrom-Portnoy, G., Lott, L.A., and Brabyn, J.A. (2014) Comparison of panel D-15 tests in a large older population. *Optometry and Vision Science* 91 (3): 284–290.

Schofield, J. (2009) Computers draw a new chapter in comics. *The Guardian*. www.theguardian.com/technology/2009/aug/12/dave-gibbons-watchmen-interview.

Science Café Little Rock (2019) Outreach. http://sciencecafelr.com/outreach.

Seddon, T. and Waterhouse, J. (2009) *Graphic Design for Non-Designers*. San Francisco, California: Chronicle Books.

Serrato Marks, G. (2018) How to make professional conferences more accessible for disabled people: guidance from actual disabled scientists. *Union of Concerned Scientists*. https://blog.ucsusa.org/science-blogger/how-to-make-professional-conferences-more-accessible-for-disabled-people-guidance-from-actual-disabled-scientists.

Serrell, B. (1982) *Making Exhibit Labels: A Step-by-Step Guide*. Nashville, Tennessee: American Association for State and Local History.

Serrell, B. (2015) *Exhibit Labels: An Interpretive Approach*. Lanham, Maryland: Rowman & Littlefield.

Shah, P. and Hoeffner, J. (2002) Review of graph comprehension research: implications for instruction. *Educational Psychology Review* 14 (1): 47–69.

Shanks, K. [@Forensictoxguy] (2019) https://twitter.com/forensictoxguy/status/1148239082021146624.

Shansky, R. [Dr Becca] (2010a) My science enemy. *Fumbling Towards Tenure Track*. www.labspaces.net/blog/528/My_Science_Enemy.

Shansky, R. [doc_becca] (2010b) www.instagram.com/p/hCfo.

Shapiro, D.W., Wenger, N.S., and Shapiro, M.F. (1994) The contributions of authors to multiauthored biomedical research papers. *JAMA* 271 (6): 438–442.

Sheffield, S. [@sarahlsheffield] (2018) https://twitter.com/sarahlsheffield/status/1026933209244856320.

Shiffman, D.S. (2012) Twitter as a tool for conservation education and outreach: what scientific conferences can do to promote live-tweeting. *Journal of Environmental Studies and Sciences* 2 (3): 257–262.

Shiffman, D.S. (2017) The benefits of Twitter for scientists. *Macroscope*. www.americanscientist.org/blog/macroscope/the-benefits-of-twitter-for-scientists.

Shiffman, D.S. [@WhySharksMatter] (2019a) https://twitter.com/WhySharks Matter/status/1154485694003175425.

Shiffman, D.S. (2019b) Welcome to the world, *Pseudobatos buthi*! *Macroscope*. www.americanscientist.org/blog/macroscope/welcome-to-the-world -pseudobatos-buthi.

Siegrist, M. (1996) The use or misuse of three-dimensional graphs to represent lower-dimensional data. *Behaviour & Information Technology* 15 (2): 96–100.

Simkin, D. and Hastie, R. (1987) An information-processing analysis of graph perception. *Journal of the American Statistical Association* 82 (398): 454–465.

Simon, M. (2019) How to be a young woman (or anything other than a grey-haired white man) at a business conference and still get shit done. https://medium. com/@morgan.simon/how-to-be-a-young-woman-or-anything-other-than -a-grey-haired-white-man-at-a-business-conference-755f0d88d182.

Simon, N.K. (2008a) Strange(r) encounters: Conditions for engagement. *Museum 2.0*. http://museumtwo.blogspot.com/2008/07/stranger-encounters -conditions-for.html.

Simon, N.K. (2008b) Exhibits and artifacts as social objects. *Museum 2.0*. http:// museumtwo.blogspot.com/2008/09/exhibits-and-artifacts-as-social.html.

Simon, N.K. (2009) The magic vest phenomenon and other wearable tools for talking to strangers. *Museum 2.0*. http://museumtwo.blogspot.com/2009/02/ magic-vest-phenomenon-and-other.html.

Simon, N.K. (2010a) When in your life were you most afraid to talk to strangers? *Museum 2.0*. http://museumtwo.blogspot.com/2010/02/when-in-your-life -were-you-most-afraid.html.

Simon, N.K. (2010b) *The Participatory Museum*. Santa Cruz, California: Museum 2.0.

Simon, N.K. (2016) *The Art of Relevance*. Santa Cruz, California: Museum 2.0.

SmartDraw Software (2019) Five tips for better flowcharts. www.smartdraw.com/ flowchart/flowchart-tips.htm.

Smith, E., Williams-Jones, B., Master, Z., Larivière, V., Sugimoto, C.R., Paul-Hus, A., Shi, M., Diller, E., Caudle, K., and Resnik, D.B. (2019) Researchers' perceptions of ethical authorship distribution in collaborative research teams. *Science and Engineering Ethics*. doi: 10.1007/s11948-019-00113-3.

Smithsonian Accessibility Program (n.d.) *Smithsonian Guidelines for Accessible Exhibition Design*. www.sifacilities.si.edu/ae_center/pdf/Accessible-Exhibition -Design.pdf.

Society for the Study of Evolution (2019) Best practices: poster presentations. www.evolutionmeetings.org/uploads/4/8/8/0/48804503/best_practices_-_ poster_and_oral_presentations.pdf.

Sokol, J. (2019) Troubled treasure. *Science* 364 (6442): 722–729. https://science. sciencemag.org/content/sci/364/6442/722.

Spear, M.E. (1952) *Charting Statistics*. New York: McGraw-Hill.

Spiegelman, A. (2002) Ballbuster. *The New Yorker*. www.newyorker.com/magazine/ 2002/07/22/ballbuster.

Stefaner, M. [@moritz_stefaner] (2019) https://twitter.com/moritz_stefaner/ status/1151524797530218499.

Stevenson, S. (2012) The greatest paper map of the United States you'll ever see. https://slate.com/culture/2012/01/the-best-american-wall-map-david-imus -the-essential-geography-of-the-united-states-of-america.html.

Stone, T.L., Adams, S., and Morioka, N. (2006) *Color Design Workbook: A Real-World Guide to Using Color in Graphic Design*. Gloucester, Massachusetts: Rockport Publishers, Inc.

Ström, M. (2016) Design better data tables. https://medium.com/mission-log/design-better-data-tables-430a30a00d8c.

Strunk, W. and White, E.B. (2000) *The Elements of Style*. Harlow: Pearson Education.

Stuart, B. [@bethstuart97] (2018) https://twitter.com/bethstuart97/status/1035898697878577152.

Tennant, J. [@Protohedgehog] (2014) https://twitter.com/Protohedgehog/status/449215667913228288.

Thompson, V.A., Turner, J.A.P., Pennycook, G., Ball, L.J., Brack, H., Ophir, Y., and Ackerman, R. (2013) The role of answer fluency and perceptual fluency as metacognitive cues for initiating analytic thinking. *Cognition* 128 (2): 237–251.

Trani, O. (2018) Blogs and social media at EGU 2018 – tune in to the conference action. *EGU Blogs*. https://blogs.egu.eu/geolog/2018/04/03/blogs-and-social-media-at-egu-2018-tune-in-to-the-conference-action.

Trost, M. (2009) The fate of the newspaper: exclusive interview with Jacek Utko. *TED Blog*. https://blog.ted.com/the_fate_of_the.

Tsybulskaya, D. and Camhi, J. (2009) Accessing and incorporating visitors' entrance narratives in guided museum tours. *Curator: the Museum Journal* 52 (1): 81–100.

Tufte, E. (1990) *Envisioning Information*. Cheshire, Connecticut: Graphics Press.

Tufte, E. (2001) *The Visual Display of Quantitative Information*. Cheshire, Connecticut: Graphics Press.

Tufte, E. (2002) Good web design, web standards, user testing. www.edwardtufte.com/bboard/q-and-a-fetch-msg?msg_id=0000P9&topic_id=1.

Tufte, E. (2003) *The Cognitive Style of PowerPoint*. Cheshire, Connecticut: Graphics Press.

Tufte, E. (2006) *Beautiful Evidence*. Cheshire, Connecticut: Graphics Press.

Van Noorden, R. (2014) Online collaboration: scientists and the social network. *Nature* 512: 126–129.

Vinkers, C.H., Tijdink, J.K., and Otte, W.M. (2015) Use of positive and negative words in scientific PubMed abstracts between 1974 and 2014: retrospective analysis. *BMJ* 351: h6467.

Visocky O'Grady, J. and Visocky O'Grady, K. (2008) *The Information Design Handbook*. Cincinatti, Ohio: HOW Books.

Vlahos, J. (2018) A designer's guide to working with CMYK black. *Printi*. www.printi.com/blog/guide-to-cmyk-black.

Wallwork, A. (2016) *English for Presentations at International Conferences*. New York: Springer.

Ware, M. (2004) Institutional repositories and scholarly publishing. *Learned Publishing* 17 (2): 115–124.

Weathers, F.W., Litz, B.T., Herman, D.S., Huska, J.A., and Keane, T.M. (1993) The PTSD Checklist (PCL): reliability, validity, and diagnostic utility. 9th International Society for Traumatic Stress Studies. San Antonio, Texas, USA.

Weissgerber, T.L., Milic, N.M., Winham, S.J., and Garovic, V.D. (2015) Beyond bar and line graphs: Time for a new data presentation paradigm. *PLoS Biology* 13 (4): e1002128.

Werne, J. (1989) At color's mercy. *Chicago Tribune*. http://articles.chicagotribune.com/1989-03-05/news/8903240191_1_color-research-carlton-wagner-wagner-institute.

Werner, K. [@kaitlynmwerner] (2016) https://twitter.com/kaitlynmwerner/status/737024318446702593.

West, L. (2012) The world according to stock photos of women. *Jezebel*. https://jezebel.com/the-world-according-to-stock-photos-of-women-5909078.

Wieman, C. (2007) The "curse of knowledge" or why intuition about teaching often fails. *American Physical Society News* 16 (10).

Wilkinson, S.E., Basto, M.Y., Perovic, G., Lawrentschuk, N., and Murphy, D.G. (2015) The social media revolution is changing the conference experience: analytics and trends from eight international meetings. *BJU International* 115 (5): 839–846.

Williams, R. (2004) *The Non-Designer's Design Book*. Berkeley, California: Peachpit Press.

Wilson, D. and Sperber, D. (2004) Relevance theory. In L.R. Horn and G. Ward (eds), *Handbook of Pragmatics*. Oxford: Blackwell, pp. 607–632.

Wilson, M.S. [@sexchrlab] (2015). https://twitter.com/sexchrlab/status/621082315758923777.

Woolson, C. (2015) Fruit-fly paper has 1,000 authors. *Nature* 521: 263.

Wurman, R.S. (2001) *Information Anxiety 2*. Indianapolis, Indiana: Que.

Yanofsky, D. (2015) It's OK not to start your y-axis at zero. *Quartz*. https://qz.com/418083/its-ok-not-to-start-your-y-axis-at-zero.

Yoder, J. [@JBYoder] (2019) https://twitter.com/JBYoder/status/1143546833622491137.

Zacks, J. and Tversky, B. (1999) Bars and lines: A study of graphic communication. *Memory and Cognition* 27 (6): 1073–1079.

Zastrow, M. (2015) Data visualization: Science on the map. *Nature* 519: 119–120.

Zimmer, C. (2019) Science writing: guidelines and guidance. https://medium.com/swlh/science-writing-guidelines-and-guidance-8c6a6bc37d75.

Zimmerman, E. (2011) Women laughing alone with salad. *The Hairpin*. www.thehairpin.com/2011/01/women-laughing-alone-with-salad.

Index

Numbers in *italics* indicate figures, and numbers in **bold** indicate tables.